Cryonics & Cryopreservation
Waiting for the Future

Contents

Chapter 1

Introduction & Overview of Cryonics

1.1 Cryonics

For the study of the production of very low temperatures, see Cryogenics. For the low-temperature preservation of living tissue and organisms in general, see Cryopreservation. For the Hot Cross album, see Cryonics (album).

Cryonics (from Greek κρύος 'kryos-' meaning 'cold') is

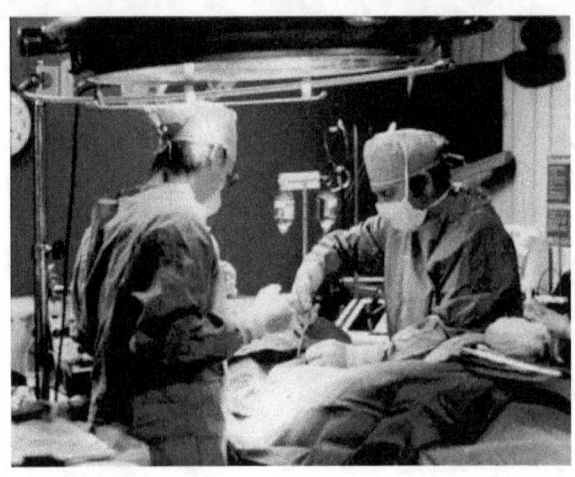

Technicians prepare a legally dead patient for cryopreservation.

the low-temperature preservation of animals and humans who cannot be sustained by contemporary medicine, with the hope that healing and resuscitation may be possible in the future.[1]

Cryopreservation of people or large animals is not reversible with current technology. The stated rationale for cryonics is that people who are considered dead by current legal or medical definitions may not necessarily be dead according to the more stringent information-theoretic definition of death.[2] It is proposed that cryopreserved people might someday be recovered by using highly advanced technology.[3]

Some scientific literature supports the feasibility of cryonics.[3][4] An open letter supporting the idea of cryonics

has been signed by 63 scientists, including Aubrey de Grey and Marvin Minsky.[5] However, many other scientists regard cryonics with skepticism.[6] As of 2014, the majority of members of the cryonics organizations are men, but the majority of those who have undergone cryopreservation procedures are women.[7]

Cryonics procedures ideally begin within minutes of cardiac arrest, and use cryoprotectants to prevent ice formation during cryopreservation.[8]

1.1.1 Premises

A central premise of cryonics is that long-term memory, personality, and identity are stored in durable cell structures and patterns within the brain that do not require continuous brain activity to survive.[9] This premise is generally accepted in medicine; it is known that under certain conditions the brain can stop functioning and still later recover with retention of long-term memory.[10] are that (1) brain structures encoding personality and long-term memory persist for some time after legal death, (2) these structures are preserved by cryopreservation, and (3) future technologies that could restore encoded memories to functional expression in a healed person are theoretically possible. At present only cells, tissues, and some small organs can be reversibly cryopreserved.[11][12]

A moral premise of cryonics is that all terminally ill patients should have the right, if they so choose, to be cryopreserved.[13]

1.1.2 Obstacles to success

Preservation injury

Long-term cryopreservation can be achieved by cooling to near 77.15 Kelvin (approximately −196.01°C), the boiling point of liquid nitrogen. It is a common mistaken belief that cells will lyse (burst) due to the formation of ice crystals within the cell, since this only occurs if the freezing rate exceeds the osmotic loss of water to the extracellular space.[14] However, damage from freezing can still be serious; ice may still form between cells, causing mechanical and chemical damage. The difficulties of recovering complex organisms from a frozen state have been long known. Attempts to recover large frozen mammals by simply rewarming were abandoned by 1957.[15]

When used at high concentrations, cryoprotectants stop ice formation completely. Cooling and solidification without crystal formation is called vitrification.[16] The first cryoprotectant solutions able to vitrify at very slow cooling rates while still being compatible with tissue survival were developed in the late 1990s by cryobiologists Gregory Fahy and Brian Wowk for the purpose of banking transplantable organs.[17][18] This has allowed animal brains to be vitrified, warmed back up, and examined for ice damage using light and electron microscopy. No ice crystal damage was found.[19]

Revival

Those who believe that revival may someday be possible generally look toward advanced bioengineering, molecular nanotechnology,[20] or nanomedicine[21] as key technologies. Revival would require repairing damage from lack of oxygen, cryoprotectant toxicity, thermal stress (fracturing), freezing in tissues that do not successfully vitrify, and reversing the effects that caused the patient's death. In many cases extensive tissue regeneration would be necessary.

It has sometimes been written that cryonics revival will be a last in, first out process. People cryopreserved in the future, with better technology, may require less advanced technology to be revived because they will have been cryopreserved with better technology that caused less damage to tissue. In this view, preservation methods would get progressively better until eventually they are demonstrably reversible, after which medicine would begin to reach back and revive people cryopreserved by more primitive methods. Revival of people cryopreserved by early cryonics technology may require centuries, if it is possible at all. The "last in, first out" view of cryonics has been criticized because the quality of cryopreservation depends on many factors other than the era in which cryopreservation takes place.[22]

1.1.3 Legal issues

Legally, cryonics patients are treated as deceased persons.[23][24] Cryonics providers tend to be treated as medical research institutes. In France, cryonics is not considered a legal mode of body disposal;[25] only burial, cremation, and formal donation to science are allowed.

However, bodies may legally be shipped to other, less restrictive countries for cryonic freezing.[26]

1.1.4 Ethical considerations

Cryonics views legal death as a perhaps sometimes pragmatically useful but fundamentally flawed and usually incorrect diagnosis which has no theoretical or philosophical justification. "Legal death" is usually just another name for a set of symptoms that have proven resistant to treatment by contemporary medicine. If death is not an event that happens suddenly when the heart stops (and "legal death" is often pronounced) this raises philosophical questions about what exactly death is. In 2005 an ethics debate in the medical journal, Critical Care, noted "...few if any patients pronounced dead by today's physicians are in fact truly dead by any scientifically rigorous criteria."[27] C Ethical and theological opinions of cryonics tend to pivot on the issue of whether cryonics is regarded as interment or medicine. If cryonics is Many followers of Nikolai Fyodorovich Fyodorov, a Russian Orthodox Christian philosopher, see cryonics as an important step in the Common Cause project which he originated.[28]

In 1969, a Roman Catholic priest consecrated the cryonics capsule of Ann DeBlasio, one of the first cryonics patients.[29]

At the request of the American Cryonics Society, in 1995, philosopher Charles Tandy, Ph.D. [30] authored a paper entitled "Cryonic-Hibernation in Light of the Bioethical Principles of Beauchamp and Childress." Tandy considered the four bioethical factors or principles articulated by philosophers Beauchamp and Childress as they apply to cryonics. These four principles are 1) respect for autonomy; 2) nonmaleficence; 3) beneficence; and 4) justice. Tandy concluded that in respect to all four principles "biomedical professionals have a strong (not weak) and actual (not prima facie, but binding) obligation to help insure cryonic-hibernation of the cryonics patient."[31]

1.1.5 History

20th century

In 1922 Alexander Yaroslavsky, member of Russian immortalists-biocosmists movement, wrote "Anabiosys Poem". However, the modern era of cryonics began in 1962 when Michigan college physics teacher Robert Ettinger proposed in a privately published book, The Prospect of Immortality,[32] that freezing people may be a way to reach future medical technology. (The book was republished in 2005 and remains in print.) Even though freezing a person is apparently fatal, Ettinger argued that what appears to be fatal today may be reversible in the future. He applied the same argument to the process of dying itself, saying that the early stages of clinical death may be reversible in the future. Combining these two ideas, he suggested that freezing recently deceased people may be a way to save lives. In 1955 James Lovelock was able to reanimate rats frozen at 0 Celsius using microwave diathermy.[33]

Slightly before Ettinger's book was complete, Evan Cooper[34] (writing as Nathan Duhring) privately published a book called *Immortality: Physically, Scientifically, Now* that independently suggested the same idea. Cooper founded the Life Extension Society (LES) in 1964 to promote freezing people. Ettinger came to be credited as the originator of cryonics, perhaps because his book was republished by Doubleday in 1964 on recommendation of Isaac Asimov and Fred Pohl, and received more publicity. Ettinger also stayed with the movement longer.

21st century

In 2015 Du Hong, a 61-year-old female writer of children's literature, became the first known Chinese person to be cryopreserved.[35]

DARPA currently funds several research projects aimed on sending the human body into a state of suspended animation, essentially "shutting down" the heart and brain until proper care can be administered that can be regarded as a step to cryopreservation of humans.[36]

1.1.6 In popular culture

Suspended animation in fiction is a popular theme in science fiction and fantasy settings, appearing in literature, comic books, films, and television. A survey in Germany found that about half of the respondents were familiar with cryonics, and about half of those familiar with cryonics had learned of the subject from television or film.[37]

Famous people

The best known cryopreserved patient is baseball player Ted Williams. The urban legend suggesting Walt Disney was cryopreserved is false; he was cremated and interred at Forest Lawn Memorial Park Cemetery.[38][39] Robert A. Heinlein, who wrote enthusiastically of the concept in *The Door into Summer*, was cremated and had his ashes distributed over the Pacific Ocean. Timothy Leary was a longtime cryonics advocate, and signed up with a major cryonics provider. He changed his mind, however, shortly before his death, and so was not cryopreserved.

Hal Finney[40] and L. Stephen Coles[41] were cryopreserved in 2014. Among cryopreserved are also James Bedford,[42] Dick Clair,[42] Thomas K. Donaldson,[42] FM-2030,[42] Jerry Leaf,[42] and John-Henry Williams.[42]

1.1.7 See also

- Chemical brain preservation
- Cryogenics
- Cryptobiosis
- Extropianism
- Hibernation
- Indefinite lifespan
- Information-theoretic death
- KrioRus
- Life extension
- Nanomedicine
- Neuropreservation
- Supercooling
- Suspended animation
- Timeship
- Vitrification in cryopreservation

1.1.8 References

[1] McKie, Robin (13 July 2002). "Cold facts about cryonics". *The Observer*. Retrieved 1 December 2013. Cryonics, which began in the Fifties, is the freezing - usually in liquid nitrogen - of human beings who have been legally declared dead. The aim of this process is to keep such individuals in a state of refrigerated limbo so that it may become possible in the future to resuscitate them, cure them of the condition that killed them, and then restore them to functioning life in an era when medical science has triumphed over the activities of the Banana Reaper.

[2] Whetstine L, Streat S, Darwin M, Crippen D (2005). "Pro/con ethics debate: When is dead really dead?". *Critical Care* **9** (6): 538–42. doi:10.1186/cc3894. PMC 1414041. PMID 16356234.

[3] Merkle RC (September 1992). "The technical feasibility of cryonics". *Medical Hypotheses* **39** (1): 6–16. doi:10.1016/0306-9877(92)90133-W. PMID 1435395. The extant literature supports but does not prove the hypothesis that cryonics is a feasible method of saving the lives of people who would otherwise certainly die.

[4] "An open letter to scientific critics of cryonics". Paul Crowley's Blog. Retrieved 2010-02-14. Though many experts in cryogenics and other relevant fields are quoted in the media as condemning cryonics practice, none have written at greater length to explain their reasons. ... this is my plea to the scientific critics of cryonics: Please criticise cryonics. If you thought that someone else had done it, if you thought that the article you'd want a cryonics hopeful to read had already been written, I hope that the surveys above show you that it really hasn't.

[5] "Scientists Open Letter on Cryonics". Retrieved 2013-03-19.

[6] Lovgren, Stefan (18 March 2005). "Corpses Frozen for Future Rebirth by Arizona Company". *National Geographic*. Retrieved 15 March 2014. Many cryobiologists, however, scoff at the idea...

[7] Bob Nelson; Kenneth Bly; Sally Magana (18 March 2014). *Freezing People Is (Not) Easy: My Adventures in Cryonics*. Lyons Press. pp. 237–. ISBN 978-1-4930-0779-0.

[8] Best BP (April 2008). "Scientific justification of cryonics practice" (PDF). *Rejuvenation Research* **11** (2): 493–503. doi:10.1089/rej.2008.0661. PMID 18321197.

[9] Mayford M, Siegelbaum SA, and Kandel ER (April 10, 2012). "Synapses and Memory Storage" (PDF). *Cold Spring Harb Perspect Biol*. doi:10.1101/cshperspect.a005751. Procedural and declarative memories differ dramatically. They use a different logic (unconscious vs. conscious recall) and they are stored in different areas of the brain. Nevertheless, these two disparate memory processes share several molecular steps and an overall molecular logic. Both are created in at least two stages: one that does not require the synthesis of new proteins and one that does. In both, short-term memory involves covalent modification of preexisting proteins and changes in the strength of preexisting synaptic connections, whereas long-term memory requires the synthesis of new proteins and the growth of new connections. Moreover, both forms of memory use PKA, mitogen-activated protein kinase (MAPK), CREB-1, and CREB-2 signaling pathways to convert short-term to long-term memory. Finally, both forms appear to use morphological changes at synapses to stabilize long-term memory

[10] Guyton, Arthur C. (1986). "The Cerebral Cortex and Intellectual Functions of the Brain". *Textbook of Medical Physiology* (7th ed.). W. B. Saunders Company. p. 658. ISBN 0-7216-1260-1. We know that secondary memory does not depend on continued activity of the nervous system, because the brain can be totally inactivated by cooling, by general anesthesia, by hypoxia, by ischemia, or by any method, and yet secondary memories that have been previously stored are still retained when the brain becomes active once again. Therefore, secondary memory must result from some actual alterations of the synapses, either physical or chemical.

[11] Fahy GM, Wowk B, Wu J (2006). "Cryopreservation of complex systems: the missing link in the regenerative medicine supply chain". *Rejuvenation Research* **9** (2): 279–91. doi:10.1089/rej.2006.9.279. PMID 16706656.

[12] Fahy GM, Wowk B, Pagotan R; et al. (July 2009). "Physical and biological aspects of renal vitrification". *Organogenesis* **5** (3): 167–75. doi:10.4161/org.5.3.9974. PMC 2781097. PMID 20046680.

[13] The Ad Hoc Committee on Medical Ethics, American College of Physicians (July 1984). "American College of Physicians Ethics Manual. Part II: Research, Other Ethical Issues. Recommended Reading". *Annals of Internal Medicine* **101** (2): 263–267. doi:10.7326/0003-4819-101-2-263. Each patient is a free agent entitled to full explanation and full decision-making authority with regard to his medical care. John Stuart Mill expressed it as: `Over himself, his own body and mind, the individual is sovereign.' The legal counterpart of patient autonomy is self-determination. Both principles deny legitimacy to paternalism by stating unequivocally that, in the last analysis, the patient determines what is right for him. ... If the [terminally ill] patient is a mentally competent adult, he has the legal right to accept or refuse any form of treatment, and his wishes must be recognized and honored by his physician.

[14] Mazur P (September 1984). "Freezing of living cells: mechanisms and implications". *The American Journal of Physiology* **247** (3 Pt 1): C125–42. PMID 6383068.

[15] Smith Audrey U (1957). "Problems in the Resuscitation of Mammals from Body Temperatures Below 0 degrees C". *Proceedings of the Royal Society of London. Series B, Biological Sciences* **147** (929): 533–44. doi:10.1098/rspb.1957.0077. JSTOR 83173.

[16] Fahy GM, MacFarlane DR, Angell CA, Meryman HT (August 1984). "Vitrification as an approach to cryopreservation". *Cryobiology* **21** (4): 407–26. doi:10.1016/0011-2240(84)90079-8. PMID 6467964.

[17] Fahy GM, Wowk B, Wu J; et al. (April 2004). "Cryopreservation of organs by vitrification: perspectives and recent advances". *Cryobiology* **48** (2): 157–78. doi:10.1016/j.cryobiol.2004.02.002. PMID 15094092.

[18] Fahy, G; Wowk, B; Wu, J; Phan, J; Rasch, C; Chang, A; Zendejas, E (2005). "Corrigendum to "Cryopreservation of organs by vitrification: perspectives and recent advances" [Cryobiology 48 (2004) 157–178]". *Cryobiology* **50** (3): 344. doi:10.1016/j.cryobiol.2005.03.002.

[19] Lemler J, Harris SB, Platt C, Huffman TM (June 2004). "The arrest of biological time as a bridge to engineered negligible senescence". *Annals of the New York Academy of Sciences* **1019** (1): 559–63. doi:10.1196/annals.1297.104. PMID 15247086.

[20] Nanofactory Collaboration http://www.MolecularAssembler.com/Nanofactory

[21] Robert A. Freitas Jr., Nanomedicine, Landes Bioscience; Vol I (1999), Vol IIA (2003) Nanomedicine.com

[22] de Wolf, Aschwin (2009-01-20). "5 dangerous idea about cryonics". Depressed Metabolism. Retrieved 2010-03-07.

[23] Legal Protection of Cryonics Patients, Part 1, INSTITUTE FOR EVIDENCE BASED CRYONICS

[24] Legal Protection of Cryonics Patients, Part 2, INSTITUTE FOR EVIDENCE BASED CRYONICS

[25] http://www.leparticulier.fr/jcms/c_101664/conseil-d-etat-du-06/01/2006-n-260307-cryogenisation-interdiction

[26] Chrisafis, Angelique (16 March 2006). "Freezer failure ends couple's hopes of life after death". *The Guardian*. Retrieved 8 January 2014.

[27] Whetstine, Leslie; Stephen Streat; Mike Darwin; David Crippen (2005-10-31). *Pro/con ethics debate: When is dead really dead?*. Critical Care Forum. Retrieved 2006-03-17.

[28] Fedorov seminar in Moscow, Russia on 25.11.2006

[29] Curtis Henderson (September–October 1969). "Cryonic Suspension of Ann DeBlasio". *Cryonics Reports* (Cryonics Society of New York, Inc.) **4** (9–10): 10–15.

[30] Tandy, Charles. "Charles Tandy, Ph.D.". Retrieved 2008-10-10.

[31] Charles Tandy (1995). "Cryonic-hibernation in light of the bioethical principles of Beauchamp and Childress". Retrieved 2008-04-01.

[32] Ettinger, Robert C.W. (1964). *The Prospect of Immortality* (First ed.). Doubleday. ISBN 0-9743472-3-X.

[33] Andjus, R.K.; Lovelock, J.E (June 28, 1955). "Reanimation of rats from body temperatures between 0 and 1C by microwave diathermy". *The Journal of Physiology* **128** (3): 541–546. PMC 1365902. PMID 13243347.

[34] "Ev Cooper". cryonet.org. Retrieved 2006-03-17.

[35] Stephen Chen. "Cheating death? Elderly writer is the first known Chinese to embrace cryogenics, her head now frozen by lab in Arizona | South China Morning Post". Scmp.com. Retrieved 2015-09-24.

[36] Put, Should We. "h+ Magazine | Covering technological, scientific, and cultural trends that are changing human beings in fundamental ways". Hplusmagazine.com. Retrieved 2012-11-18.

[37] Kaiser S1, Gross D1, Lohmeier J1, Rosentreter M1, Raschke J2 (2014). "Attitudes and acceptance toward the technology of cryonics in Germany". *INTERNATIONAL JOURNAL OF TECHNOLOGY ASSESSMENT IN HEALTH CARE* **5** (1): 1–7. doi:10.1017/S0266462313000718. PMID 24499638.

[38] "Urban Legends Reference Pages: Disney (Suspended Animation)". *snopes.com*.

[39] "The Truth About Walt Disney and Cryogenics". *mentalfloss.com*.

[40] "Bitcoin's Earliest Adopter Is Cryonically Freezing His Body to See the Future - WIRED". *WIRED*.

[41] Los Angeles Times (4 December 2014). "L. Stephen Coles dies at 73; studied extreme aging in humans". *latimes.com*.

[42] "The Quick 8: Eight People Who Have Been Cryonically Preserved (and one who wasn't)". Mental Floss. 2009-02-11. Retrieved 2015-09-13.

1.1.9 External links

- Cryonics at DMOZ

1.2 Cryopreservation

Cryopreservation of plant shoots. Open tank of liquid nitrogen behind.

Cryopreservation or **cryoconservation** is a process where cells, whole tissues, or any other substances susceptible

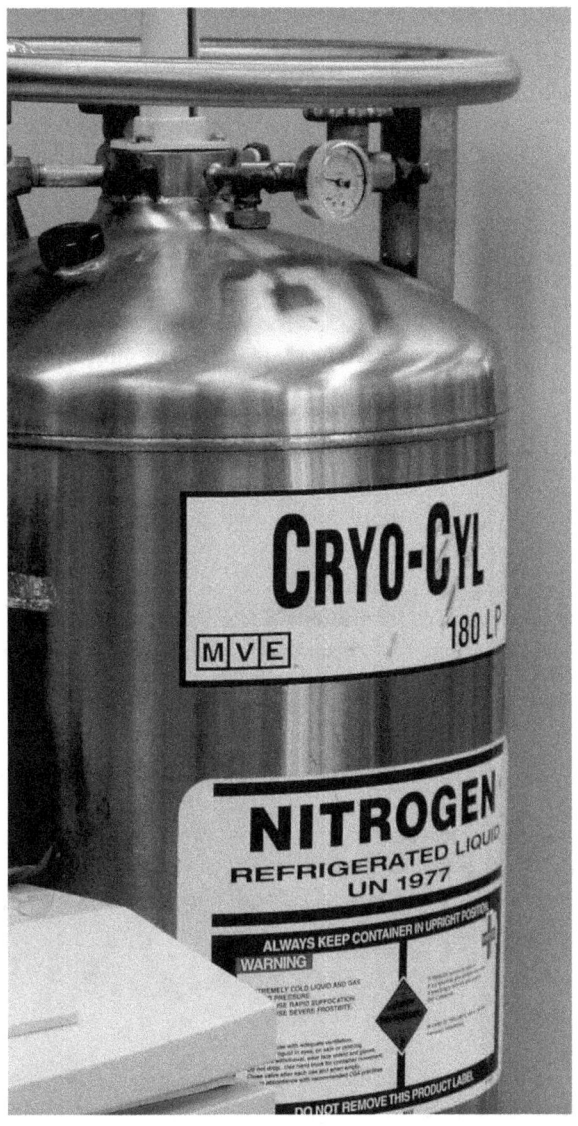

A tank of liquid nitrogen, used to supply a cryonic freezer (for storing laboratory samples at a temperature of about −150 °C)

to damage caused by chemical reactivity or time are preserved by cooling to sub-zero temperatures. At low enough temperatures, any enzymatic or chemical activity which might cause damage to the material in question is effectively stopped. Cryopreservation methods seek to reach low temperatures without causing additional damage caused by the formation of ice during freezing. Traditional cryopreservation has relied on coating the material to be frozen with a class of molecules termed cryoprotectants. New methods are constantly being investigated due to the inherent toxicity of many cryoprotectants. By default it should be considered that cryopreservation alters or compromises the structure and function of cells unless it is proven otherwise for a particular cell population.

1.2.1 Natural cryopreservation

Water-bears (Tardigrada), microscopic multicellular organisms, can survive freezing by replacing most of their internal water with the sugar trehalose, preventing it from crystallization that otherwise damages cell membranes. Mixtures of solutes can achieve similar effects. Some solutes, including salts, have the disadvantage that they may be toxic at intense concentrations. In addition to the water-bear, wood frogs can tolerate the freezing of their blood and other tissues. Urea is accumulated in tissues in preparation for overwintering, and liver glycogen is converted in large quantities to glucose in response to internal ice formation. Both urea and glucose act as "cryoprotectants" to limit the amount of ice that forms and to reduce osmotic shrinkage of cells. Frogs can survive many freeze/thaw events during winter if not more than about 65% of the total body water freezes. Research exploring the phenomenon of "Freezing frogs" has been performed primarily by the Canadian researcher, Dr. Kenneth B. Storey.

Freeze tolerance, in which organisms survive the winter by freezing solid and ceasing life functions, is known in a few vertebrates: five species of frogs (*Rana sylvatica*, *Pseudacris triseriata*, *Hyla crucifer*, *Hyla versicolor*, *Hyla chrysoscelis*), one of salamanders (*Hynobius keyserlingi*), one of snakes (*Thamnophis sirtalis*) and three of turtles (*Chrysemys picta*, *Terrapene carolina*, *Terrapene ornata*).[1] Snapping turtles *Chelydra serpentina* and wall lizards *Podarcis muralis* also survive nominal freezing but it has not been established to be adaptive for overwintering. In the case of *Rana sylvatica* one cryopreservant is ordinary glucose, which increases in concentration by approximately 19 mmol/l when the frogs are cooled slowly.[1]

1.2.2 History

One of the most important early theoreticians of cryopreservation was James Lovelock (born 1919) of Gaia theory fame. He suggested that damage to red blood cells during freezing was due to osmotic stress. Lovelock during the early 1950s had also suggested that increasing salt concentrations in a cell as it dehydrates to lose water to the external ice might cause damage to the cell.[2] Cryopreservation of tissue during recent times began with the freezing of fowl sperm, which during 1957 was cryopreserved by a team of scientists in the UK directed by Christopher Polge.[3] The process was applied to humans during the 1950s with pregnancies obtained after insemination of frozen sperm. However, the rapid immersion of the samples in liquid nitrogen did not, for certain of these samples – such as types of embryos, bone marrow and stem cells – produce the necessary viability to make them usable after thawing. Increased understanding of the mechanism of freezing injury to cells emphasised the importance of controlled or slow cooling to obtain maximum survival on thawing of the living cells. A controlled-rate cooling process, allowing biological samples to equilibrate to optimal physical parameters osmotically in a cryoprotectant (a form of anti-freeze) before cooling in a predetermined, controlled way proved necessary. The ability of cryoprotectants, in the early cases glycerol, to protect cells from freezing injury was discovered accidentally. Freezing injury has two aspects: direct damage from the ice crystals and secondary damage caused by the increase in concentration of solutes as progressively more ice is formed. During 1963 Peter Mazur, at Oak Ridge National Laboratory in the USA, demonstrated that lethal intracellular freezing could be avoided if cooling was slow enough to permit sufficient water to leave the cell during progressive freezing of the extracellular fluid. That rate differs between cells of differing size and water permeability: a typical cooling rate around 1 °C/minute is appropriate for many mammalian cells after treatment with cryoprotectants such as glycerol or dimethyl sulphoxide, but the rate is not a universal optimum.

1.2.3 Temperature

Cryonic storage at very cold temperatures is presumed to provide an indefinite longevity to cells, although the actual effective life is rather difficult to prove. Researchers experimenting with dried seeds found that there was noticeable variability of deterioration when samples were kept at different temperatures – even ultra-cold temperatures. Temperatures less than the glass transition point (Tg) of polyol's water solutions, around −136 °C (137 K; −213 °F), seem to be accepted as the range where biological activity very substantially slows, and −196 °C (77 K; −321 °F), the boiling point of liquid nitrogen, is the preferred temperature for storing important specimens. While refrigerators, freezers and extra-cold freezers are used for many items, generally the ultra-cold of liquid nitrogen is required for successful preservation of the more complex biological structures to virtually stop all biological activity.

1.2.4 Risks

Phenomena which can cause damage to cells during cryopreservation mainly occur during the freezing stage, and include: solution effects, extracellular ice formation, dehydration and intracellular ice formation. Many of these effects can be reduced by cryoprotectants. Once the preserved material has become frozen, it is relatively safe from further damage. However, estimates based on the accumulation of radiation-induced DNA damage during cryonic storage have suggested a maximum storage period of 1000 years.[4]

Solution effects As ice crystals grow in freezing water, solutes are excluded, causing them to become concentrated in the remaining liquid water. High concentrations of some solutes can be very damaging.

Extracellular ice formation When tissues are cooled slowly, water migrates out of cells and ice forms in the extracellular space. Too much extracellular ice can cause mechanical damage to the cell membrane due to crushing.

Dehydration Migration of water, causing extracellular ice formation, can also cause cellular dehydration. The associated stresses on the cell can cause damage directly.

Intracellular ice formation While some organisms and tissues can tolerate some extracellular ice, any appreciable intracellular ice is almost always fatal to cells.

1.2.5 Main methods to prevent risks

The main techniques to prevent cryopreservation damages are a well established combination of *controlled rate and slow freezing* and a newer flash-freezing process known as *vitrification*.

Slow programmable freezing

Controlled-rate and slow freezing, also known as *slow programmable freezing (SPF)*,[5] is a set of well established techniques developed during the early 1970s which enabled the first human embryo frozen birth Zoe Leyland during 1984. Since then, machines that freeze biological samples using programmable sequences, or controlled rates, have been used all over the world for human, animal and cell biology – "freezing down" a sample to better preserve it for eventual thawing, before it is frozen, or cryopreserved, in liquid nitrogen. Such machines are used for freezing oocytes, skin, blood products, embryo, sperm, stem cells and general tissue preservation in hospitals, veterinary practices and research laboratories around the world. As an example, the number of live births from frozen embryos 'slow frozen' is estimated at some 300,000 to 400,000 or 20% of the estimated 3 million in vitro fertilisation (IVF) births.[6]

Lethal intracellular freezing can be avoided if cooling is slow enough to permit sufficient water to leave the cell during progressive freezing of the extracellular fluid. That rate differs between cells of differing size and water permeability: a typical cooling rate of about 1 °C/minute is appropriate for many mammalian cells after treatment with cryoprotectants such as glycerol or dimethyl sulphoxide, but the rate is not a universal optimum. The 1 °C / minute rate can be achieved by using devices such as a rate-controlled freezer or a benchtop portable freezing container.[7]

Several independent studies have provided evidence that frozen embryos stored using slow-freezing techniques may in some ways be 'better' than fresh in IVF. The studies were presented at the American Society for Reproductive Medicine conference in San Francisco, USA, 2008. The studies indicate that using frozen embryos and eggs rather than fresh embryos and eggs reduced the risk of stillbirth and premature delivery though the exact reasons are still being explored.[8]

Vitrification

Researchers Greg Fahy and William F. Rall helped introduce vitrification to reproductive cryopreservation in the mid-1980s.[9] As of 2000, researchers claim vitrification provides the benefits of cryopreservation without damage due to ice crystal formation.[10] For clinical cryopreservation, vitrification usually requires the addition of cryoprotectants prior to cooling. The cryoprotectants act like antifreeze: they decrease the freezing temperature. They also increase the viscosity. Instead of crystallizing, the syrupy solution becomes an amorphous ice—it *vitrifies*. Rather than a phase change from liquid to solid by crystallization, the amorphous state is like a "solid liquid", and the transformation is over a small temperature range described as the "glass transition" temperature.

Vitrification of water is promoted by rapid cooling, and can be achieved without cryoprotectants by an extremely rapid decrease of temperature (megakelvins per second). The rate that is required to attain glassy state in pure water was considered to be impossible until 2005.[11]

Two conditions usually required to allow vitrification are an increase of the viscosity and a decrease of the freezing temperature. Many solutes do both, but larger molecules generally have larger effect, particularly on viscosity. Rapid cooling also promotes vitrification.

For established methods of cryopreservation, the solute must penetrate the cell membrane in order to achieve increased viscosity and decrease freezing temperature inside the cell. Sugars do not readily permeate through the membrane. Those solutes that do, such as dimethyl sulfoxide, a common cryoprotectant, are often toxic in intense concentration. One of the difficult compromises of vitrifying cryopreservation concerns limiting the damage produced by the cryoprotectant itself due to cryoprotectant toxicity. Mixtures of cryoprotectants and the use of ice blockers have enabled the Twenty-First Century Medicine company to vitrify a rabbit kidney to −135 °C with their proprietary vitrification mixture. Upon rewarming, the kidney was transplanted successfully into a rabbit, with complete functionality and viability, able to sustain the rabbit indefinitely as the sole functioning kidney.[12]

Cells Alive System freezers

The makers of Cells Alive System, a programmable freezer originally designed for food storage, claim that the survival of conventionally frozen periodontal ligament cells is improved by the application of a combination of a 0.1 mT magnetic field oscillating at 60 Hz, an induced electric field, and mechanical and thermal vibrations.[13][14][15][16][17][18]

1.2.6   Freezable tissues

Generally, cryopreservation is easier for thin samples and small clumps of individual cells, because these can be cooled more quickly and so require lesser doses of toxic cryoprotectants. Therefore, cryopreservation of human livers and hearts for storage and transplant is still impractical.

Nevertheless, suitable combinations of cryoprotectants and regimes of cooling and rinsing during warming often allow the successful cryopreservation of biological materials, particularly cell suspensions or thin tissue samples. Examples include:

- Semen in semen cryopreservation

- Blood

    - Special cells for transfusion

    - Stem cells. It is optimal in high concentration of synthetic serum, stepwise equilibration and slow cooling.[19]

    - Umbilical cord blood *Further information: Cord blood bank#Cryopreservation*

- Tissue samples like tumors and histological cross sections

- Eggs (oocytes) in oocyte cryopreservation

- Embryos at cleavage stage (that are 2, 4 or 8 cells) or at blastocyst stage, in embryo cryopreservation

- Ovarian tissue in ovarian tissue cryopreservation

- Plant seeds or shoots may be cryopreserved for conservation purposes.

Additionally, efforts are underway to preserve humans cryogenically, known as cryonics. For such efforts either the brain within the head or the entire body may experience the above process. Cryonics is in a different category from the aforementioned examples, however: while countless cryopreserved cells, vaccines, tissue and other biologial samples have been thawed and used successfully, this has not yet been the case at all for cryopreserved brains or bodies. At issue are the criteria for defining "success". Proponents of cryonics claim that cryopreservation using present technology, particularly vitrification of the brain, may be sufficient to preserve people in an "information theoretic" sense so that they could be revived and made whole by hypothetical vastly advanced future technology.

Embryos

Main article: Embryo cryopreservation

Cryopreservation for embryos are used for *embryo storage*, e.g. when in vitro fertilization has resulted in more embryos than is currently needed.

Pregnancies have been reported from embryos stored for 16 years.[20] Many studies have evaluated the children born from frozen embryos, or "frosties". The result has been uniformly positive with no increase of birth defects or development abnormalities.[21] A study of more than 11,000 cryopreserved human embryos had no significant effect of storage time on post-thaw survival for in vitro fertilisation (IVF) or oocyte donation cycles, or for embryos frozen at the pronuclear or cleavage stages.[22] Additionally, the duration of storage did not have any significant effect on clinical pregnancy, miscarriage, implantation, or live birth rate, whether from IVF or oocyte donation cycles.[22] Rather, oocyte age, survival proportion, and number of transferred embryos are predictors of pregnancy outcome.[22]

Ovarian tissue

Main article: Ovarian tissue cryopreservation

Cryopreservation of ovarian tissue is of interest to women who want to preserve their reproductive function beyond the natural limit, or whose reproductive potential is threatened by cancer therapy,[23] for example in hematologic malignancies or breast cancer.[24] The procedure is to take a part of the ovary and perform slow freezing before storing it in liquid nitrogen whilst therapy is undertaken. Tissue can then be thawed and implanted near the fallopian, either orthotopic (on the natural location) or heterotopic (on the abdominal wall),[24] where it starts to produce new eggs, allowing normal conception to occur.[25] The ovarian tissue may also be transplanted into mice that are immunocompromised (SCID mice) to avoid graft rejection, and tissue can be harvested later when mature follicles have developed.[26]

Oocytes

Main article: Oocyte cryopreservation

Human *Oocyte cryopreservation* is a new technology in which a woman's eggs (oocytes) are extracted, frozen and stored. Later, when she is ready to become pregnant, the eggs can be thawed, fertilized, and transferred to the uterus as embryos.

Semen

Main article: Semen cryopreservation

Semen can be used successfully almost indefinitely after cryopreservation. The longest reported successful storage is 22 years.[27] It can be used for sperm donation where the recipient wants the treatment in a different time or place, or as a means of preserving fertility for men undergoing vasectomy or treatments that may compromise their fertility, such as chemotherapy, radiation therapy or surgery.

Testicular tissue

Cryopreservation of immature testicular tissue is a developing method to avail reproduction to young boys who need to have gonadotoxic therapy. Animal data are promising, since healthy offsprings have been obtained after transplantation of frozen testicular cell suspensions or tissue pieces. However, none of the fertility restoration options from frozen tissue, i.e. cell suspension transplantation, tissue grafting and in vitro maturation (IVM) has proved efficient and safe in humans as yet.[28]

Moss

Cryopreservation of whole moss plants, especially Physcomitrella patens, has been developed by Ralf Reski and coworkers[29] and is performed at the International Moss Stock Center. This biobank collects, preserves, and distributes moss mutants and moss ecotypes.[30]

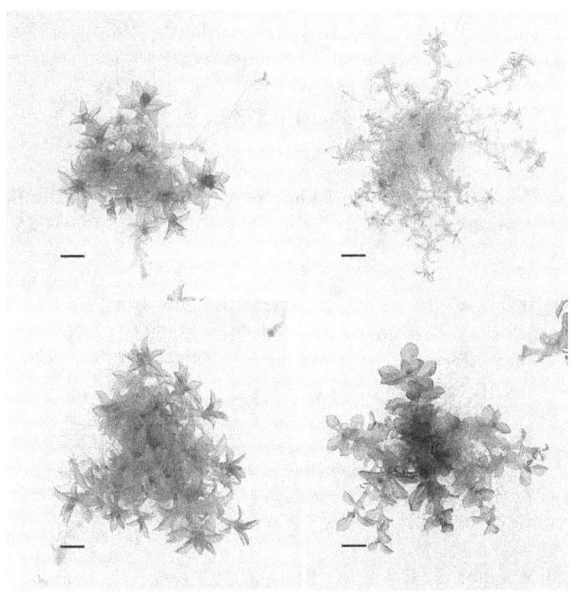

Four different ecotypes of Physcomitrella patens *stored at the IMSC.*

Mesenchymal stromal cells (MSCs)

Scientists have reported that MSCs when transfused immediately within few hours post thawing may show reduced function or show decreased efficacy in treating diseases as compared to those MSCs which are in log phase of cell growth(fresh), so cryopreserved MSCs should be brought back into log phase of cell growth in *in vitro* culture before these are administered for clinical trials or experimental therapies, re-culturing of MSCs will help in recovering from the shock the cells get during freezing and thawing. Various clinical trials on MSCs have failed which used cryopreserved product immediately post thaw as compared to those clinical trials which used fresh MSCs [Francois M et al., Cytotherapy.2012;14:147–152].

1.2.7   Preservation of microbiology cultures

Bacteria and fungi can be kept short-term (months to about a year, depending) refrigerated, however, cell division and metabolism is not completely arrested and thus is not an optimal option for long-term storage (years) or to preserve cultures genetically or phenotypically, as cell divisions can lead to mutations or sub-culturing can cause phenotypic changes. A preferred option, species-dependent, is cryopreservation.

Fungi

Fungi, notably zygomycetes, ascomycetes and higher basidiomycetes, regardless of sporulation, are able to be stored in liquid nitrogen or deep-frozen. Crypreservation is a hall-mark method for fungi that do not sporulate (otherwise other preservation methods for spores can be used at lower costs and ease), sporulate but have delicate spores (large or freeze-dry sensitive), are pathogenic (dangerous to keep metabolically active fungus) or are to be used for genetic stocks (ideally to have identical composition as the original deposit). As with many other organisms, cryoprotectants like DMSO or glycerol (e.g. filamentous fungi 10% glycerol or yeast 20% glycerol) are used. Differences between choosing cryoprotectants are species (or class) dependent, but generally for fungi penetrating cryoprotectants like DMSO, glycerol or polyethylene glycol are most effective (other non-penetrating ones include sugars mannitol, sorbitol, dextran, etc.). Freeze-thaw repetition is not recommended as it can decrease viability. Back-up deep-freezers or liquid nitrogen storage sites are recommended. Multiple protocols for freezing are summarized below (each uses screw-cap polypropylene cryotubes):[31]

A) Non-sporulating fungi or embedded mycelia: 10% glycerol is added to the tube and agar plugs of fresh culture are added and immediately frozen in liquid-nitrogen vapor (−170 °C). Cultures are thawed at 37 °C and plated.

B) Spores or mycelia from agar plate: 10% glycerol or 5% DMSO spore or mycelia suspension are made and frozen.

C) Liquid mycelia: Mycelia are macerated (not for use with human pathogenic fungi) and mixed to make a final concentration of 10% glycerol or 5% DSMO.

For protocol B and C, stocks are gradually cooled until frozen. Similar thawing and plating as in A.

Bacteria

Many common culturable laboratory strains are deep-frozen to preserve genetically and phenotypically stable, long-term stocks. Sub-culturing and prolonged refrigerated samples may lead to loss of plasmid(s) or mutations. Common final glycerol percentages are 15, 20 and 25. From a fresh culture plate, one single colony of interest is chosen and liquid culture is made. From the liquid culture, the medium is directly mixed with equal amount of glycerol; the colony should be checked for any defects like mutations. All antibiotics should be washed from the culture before long-term storage. Methods vary, but mixing can be done gently by inversion or rapidly by vortex and cooling can vary by either placing the cryotube directly at −80 °C, shock-freezing in liquid nitrogen or gradually cooling and then storing at −80 °C or cooler (liquid nitrogen or liquid nitrogen vapor). Recovery of bacteria can also vary, namely if beads are stored within the tube then the few beads can be used to plate or the frozen stock can be scraped with a loop and then plated, however, since only little stock is needed the entire tube should never be completely thawed and repeated freeze-thaw should be avoided. 100% recovery is not feasible regardless of methodology.[32][33][34]

1.2.8 See also

- Cells Alive System freezers

- Chemical brain preservation

- Cryobiology

- Cryogenics

- Cryonics

- Cryostasis (clathrate hydrates)

- Ex-situ conservation

- Frozen zoo

- Futurama

1.2.9 Notes

[1] Jon P. Costanzo; Richard E. Lee; Michael F. Wright (1991). "Glucose loading prevents freezing injury in rapidly cooled wood frogs" (PDF). *American Journal of Physiology*: R1549–R1553.

[2] Mazur P (1970). "Cryobiology: the freezing of biological systems". *Science* **168** (3934): 939–49. Bibcode:1970Sci...168..939M. doi:10.1126/science.168.3934.939. PMID 5462399.

[3] Polge C (1957). "Low-Temperature Storage of Mammalian Spermatozoa". *Royal Society of London* **147** (929): 498–508. Bibcode:1957RSPSB.147..498P. doi:10.1098/rspb.1957.0068.

[4] Mazur P (1984). "Freezing of living cells: mechanisms and implications". *American Journal of Physiology* **247** (3 Pt 1): C125–42. Bibcode:1957RSPSB.147..498P. doi:10.1098/rspb.1957.0068.

[5] Vutyavanich T; Piromlertamorn W; Nunta S (April 2010). "Rapid freezing versus slow programmable freezing of human spermatozoa". *Fertil. Steril.* **93** (6): 1921–8. doi:10.1016/j.fertnstert.2008.04.076. PMID 19243759.

[6] http://www.bionews.org.uk/commentary.lasso?storyid= 4055. Missing or empty |title= (help)

[7] Thompson M; Nemits M; Ehrhardt R (May 2011). "Rate-controlled Cryopreservation and Thawing of Mammalian Cells". *Nat. Prot. Exch.* doi:10.1038/protex.2011.224.

[8] "Fresh vs. Frozen Donor Eggs". *Pacific Fertility Egg Bank*. Retrieved 9 March 2015.

[9] Rall, WF; Fahy, GM (Feb 14–20, 1985). "Ice-free cryopreservation of mouse embryos at −196 degrees C by vitrification". *Nature* **313** (6003): 573–5. Bibcode:1985Natur.313..573R. doi:10.1038/313573a0. PMID 3969158.

[10] "Alcor: The Origin of Our Name" (PDF). Alcor Life Extension Foundation. Winter 2000. Retrieved 2009-08-25.

[11] Bhat SN; Sharma A; Bhat SV (2005). "Vitrification and glass transition of water: insights from spin probe ESR". *Phys Rev Lett* **95** (23): 235702. arXiv:cond-mat/0409440. Bibcode:2005PhRvL..95w5702B. doi:10.1103/PhysRevLett.95.235702. PMID 16384318.

[12] Fahy GM; Wowk B; Pagotan R; Chang A; et al. (2009). "Physical and biological aspects of renal vitrification". *Organogenesis* **5** (3): 167–175. doi:10.4161/org.5.3.9974. PMC 2781097. PMID 20046680.

[13] Abedini, S.; Kaku, M.; Kawata, T.; Koseki, H.; Kojima, S.; Sumi, H.; Motokawa, M.; Fujita, T.; Ohtani, J.; Ohwada, N.; Tanne, K. (2011). "Effects of cryopreservation with a newly-developed magnetic field programmed freezer on periodontal ligament cells and pulp tissues". *Cryobiology* **62** (3): 181–187. doi:10.1016/j.cryobiol.2011.03.001. PMID 21397593.

[14] Kaku, M.; Kamada, H.; Kawata, T.; Koseki, H.; Abedini, S.; Kojima, S.; Motokawa, M.; Fujita, T.; Ohtani, J.; Tsuka, N.; Matsuda, Y.; Sunagawa, H.; Hernandes, R. A. M.; Ohwada, N.; Tanne, K. (2010). "Cryopreservation of periodontal ligament cells with magnetic field for tooth banking". *Cryobiology* **61** (1): 73–78. doi:10.1016/j.cryobiol.2010.05.003. PMID 20478291.

[15] Kelly, Tim (June 2, 2008). "Mr. Freeze—Norio Owada's freezing method can keep milk fresh for months. Livers, too.". *Forbes*. Retrieved September 21, 2012.

[16] Owada, N. (2009) "Quick Freezing Apparatus and Quick Freezing Method" *US Patent Application*

[17] Owada, N. (2002) "Highly-efficient freezing apparatus and highly-efficient freezing method" *US Patent 7,237,400*

[18] Wowk, B. (2012). "Electric and magnetic fields in cryopreservation". *Cryobiology* **64** (3): 301–3; author's response: 304–5. doi:10.1016/j.cryobiol.2012.02.003. PMID 22330639.

[19] Lee JY; Lee JE; Kim DK; Yoon TK; et al. (November 2008). "High concentration of synthetic serum, stepwise equilibration and slow cooling as an efficient technique for large-scale cryopreservation of human embryonic stem cells". *Fertil. Steril.* **93** (3): 976–85. doi:10.1016/j.fertnstert.2008.10.017. PMID 19022437.

[20] Planer NEWS and Press Releases > 'Twins' born 16 years apart. 01/06/2006

[21] "Genetics & IVF Institute". Givf.com. Archived from the original on 2009-07-27. Retrieved 2009-07-27.

[22] Riggs R; Mayer J; Dowling-Lacey D; Chi TF; et al. (November 2008). "Does storage time influence postthaw survival and pregnancy outcome? An analysis of 11,768 cryopreserved human embryos". *Fertil. Steril.* **93** (1): 109–15. doi:10.1016/j.fertnstert.2008.09.084. PMID 19027110.

[23] Isachenko V; Lapidus I; Isachenko E; et al. (2009). "Human ovarian tissue vitrification versus conventional freezing: morphological, endocrinological, and molecular biological evaluation". *Reproduction* **138** (2): 319–27. doi:10.1530/REP-09-0039. PMID 19439559.

[24] Oktay K; Oktem O (November 2008). "Ovarian cryopreservation and transplantation for fertility preservation for medical indications: report of an ongoing experience". *Fertil. Steril.* **93** (3): 762–8. doi:10.1016/j.fertnstert.2008.10.006. PMID 19013568.

[25] Livebirth after orthotopic transplantation of cryopreserved ovarian tissue The Lancet, September 24, 2004

[26] Lan C; Xiao W; Xiao-Hui D; Chun-Yan H; et al. (December 2008). "Tissue culture before transplantation of frozen-thawed human fetal ovarian tissue into immunodeficient mice". *Fertil. Steril.* **93** (3): 913–9. doi:10.1016/j.fertnstert.2008.10.020. PMID 19108826.

[27] Planer NEWS and Press Releases > Child born after 22 year semen storage using Planer controlled rate freezer 14/10/2004

[28] Wyns C; Curaba M; Vanabelle B; Van Langendonckt A; et al. (2010). "Options for fertility preservation in prepubertal boys". *Hum. Reprod. Update* **16** (3): 312–28. doi:10.1093/humupd/dmp054. PMID 20047952.

[29] Schulte, J., Ralf Reski (2004): High-throughput cryopreservation of 140000 Physcomitrella patens mutants. Plant Biol. 6, 119-127. Schulte J.; Reski R. "High throughput cryopreservation of 140,000 Physcomitrella patens mutants". Plant Biotechnology, Freiburg University, Freiburg, Germany. Retrieved 17 August 2010.

[30] ScienceDaily: Mosses, deep frozen. "Mosses, deep-frozen".

[31] http://128.104.77.228/documnts/pdf2004/fpl_2004_nakasone001.pdf

[32] Freeze-Drying and Cryopreservation of Bacteria

[33] "Addgene: Protocol - How to Create a Bacterial Glycerol Stock". *addgene.org*. Retrieved 9 September 2015.

[34] http://www.qiagen.com/knowledge-and-support/spotlight/plasmid-resource-center/growth%20of%20bacterial%20cultures/

1.2.10 References

- Engelmann, F.; M. E. Dulloo; C. Astorga; S. Dussert; F. Anthony, eds. (2007). *Conserving coffee genetic resources*. Bioversity International, CATIE, IRD. p. 61.

- Panis, B & Tien Thinh, N. (2001). *Cryopreservation of Musa germplasm*. INIBAP (now Bioversity International). p. 45.

- ReproTech Limited (2012). "Fertility Preservation". ReproTech Limited.

- Nakasone, Karen K, et al., 2004, PRESERVATION AND DISTRIBUTION OF FUNGAL CULTURES.

- Stephen, F., 1995, Freeze-Drying and Cryopreservation of Bacteria

1.2.11 External links

- Vitrification for storage of embryos, HFEA website
- The Freezing of Human Oocytes (Eggs)
- Society for Cryobiology
- The Society for Low Temperature Biology
- Cellular cryobiology and anhydrobiology
- Death in the Deep Freeze
- In vitro storage and cryopreservation
- Cryonics

Chapter 2

Cryonics & Cryopreservation

2.1 Biostasis

Biostasis is the ability of an organism to tolerate environmental changes without having to actively adapt to them. The word is also used as a synonym for cryostasis or cryonics. It is found in organisms that live in habitats that may encounter unfavourable living conditions (i.e. drought, freezing, a change in pH, pressure, or temperature). Insects undergo diapause, which allows them to survive winter and other events. Diapause may be obligatory (required for the insect to survive) or facultative (the insect is able to undergo change before the initiating event arrives).

Medical Biostasis can be put to use in humans to help repair brain damage. [1]

Depending on where medicine is in the next decade medical biostasis procedures can be performed by trauma surgeons by 2026.[2]

2.1.1 References

[1] http://biostasis.businesscatalyst.com/science.html

[2] http://biostasis.businesscatalyst.com/

2.1.2 External links

- Tardigrade Facts
- Medical Biostasis Inc

2.2 Cryobiology

Cryobiology is the branch of biology that studies the effects of low temperatures on living things within Earth's cryosphere or in science. The word cryobiology is derived from the Greek words κρῦος [kryos], « cold », βίος [bios], « life », and λόγος [logos], « word » (hence science). In practice, cryobiology is the study of biological material or systems at temperatures below normal. Materials or systems studied may include proteins, cells, tissues, organs, or whole organisms. Temperatures may range from moderately hypothermic conditions to cryogenic temperatures.

2.2.1 Areas of study

At least six major areas of cryobiology can be identified: 1) study of cold-adaptation of microorganisms, plants (cold hardiness), and animals, both invertebrates and vertebrates (including hibernation), 2) cryopreservation of cells, tissues, gametes, and embryos of animal and human origin for (medical) purposes of long-term storage by cooling to temperatures below the freezing point of water. This usually requires the addition of substances which protect the cells during freezing and thawing (cryoprotectants), 3) preservation of organs under hypothermic conditions for transplantation, 4) lyophilization (freeze-drying) of pharmaceuticals, 5) cryosurgery, a (minimally) invasive approach for the destruction of unhealthy tissue using cryogenic gases/fluids, and 6) physics of supercooling, ice nucleation/growth and mechanical engineering aspects of heat transfer during cooling and warming, as applied to biological systems. Cryobiology would include cryonics, the low temperature preservation of humans and mammals with the intention of future revival, although this is not part of mainstream cryobiology, depending heavily on speculative technology yet to be invented. Several of these areas of study rely on cryogenics, the branch of physics and engineering that studies the production and use of very low temperatures

2.2.2 Cryopreservation in nature

Many living organisms are able to tolerate prolonged periods of time at temperatures below the freezing point of water. Most living organisms accumulate cryoprotectants such as antinucleating proteins, polyols, and glucose to protect themselves against frost damage by sharp ice crystals. Most plants, in particular, can safely reach temperatures of

−4 °C to −12 °C.

Bacteria

Three species of bacteria, *Carnobacterium pleistocenium*, *Chryseobacterium greenlandensis.* and *Herminiimonas glaciei*, have reportedly been revived after surviving for thousands of years frozen in ice. Certain bacteria, notably *Pseudomonas syringae*, produce specialized proteins that serve as potent ice nucleators, which they use to force ice formation on the surface of various fruits and plants at about −2 °C.[1] The freezing causes injuries in the epithelia and makes the nutrients in the underlying plant tissues available to the bacteria.[2] *Listeria* grows slowly in temperatures as low as − 1.5 °C and persists for some time in frozen foods.[3]

Plants

Many plants undergo a process called hardening which allows them to survive temperatures below 0 °C for weeks to months.

Animals

Invertebrates   Nematodes that survive below 0 °C include *Trichostrongylus colubriformis* and *Panagrolaimus davidi*. Cockroach nymphs (*Periplaneta japonica*) survive short periods of freezing at −6 to −8 °C. The red flat bark beetle (*Cucujus clavipes*) can survive after being frozen to −150 °C.[4] The fungus gnat *Exechia nugatoria* can survive after being frozen to −50 °C, by a unique mechanism whereby ice crystals form in the body but not the head. Another freeze-tolerant beetle is *Upis ceramboides*.[5] See insect winter ecology and antifreeze protein. Another invertebrate that is briefly tolerant to temperatures down to −273 °C is the tardigrade.

The larvae of *Haemonchus contortus*, a nematode, can survive 44 weeks frozen at −196 °C.

Vertebrates   For the wood frog (*Rana sylvatica*), in the winter, as much as 45% of its body may freeze and turn to ice. "Ice crystals form beneath the skin and become interspersed among the body's skeletal muscles. During the freeze, the frog's breathing, blood flow, and heartbeat cease. Freezing is made possible by specialized proteins and glucose, which prevent intracellular freezing and dehydration." [6][7] The wood frog can survive up to 11 days frozen at −4 °C.

Other vertebrates that survive at body temperatures below 0 °C include painted turtles (*Chrysemys picta*), gray tree frogs (*Hyla versicolor*), box turtles (*Terrapene carolina* -

48 hours at −2 °C), spring peeper (*Pseudacris crucifer*), garter snakes (*Thamnophis sirtalis*- 24 hours at −1.5 °C), the chorus frog (*Pseudacris triseriata*), Siberian salamander (*Salamandrella keyserlingii* - 24 hours at −15.3 °C),[8] European common lizard (*Lacerta vivipara*) and Antarctic fish such as *Pagothenia borchgrevinki*.[9][10] Antifreeze proteins cloned from such fish have been used to confer frost-resistance on transgenic plants.

Hibernating Arctic ground squirrels may have abdominal temperatures as low as −2.9 °C (26.8 °F), maintaining subzero abdominal temperatures for more than three weeks at a time, although the temperatures at the head and neck remain at 0 °C or above.[11]

2.2.3 Applied cryobiology

Historical background

Boyle

Cryobiology history can be traced back to antiquity. As early as in 2500 BC, low temperatures were used in Egypt in medicine. The use of cold was recommended by Hippocrates to stop bleeding and swelling. With the emergence of modern science, Robert Boyle studied the effects of low temperatures on animals.

In 1949, bull semen was cryopreserved for the first time by a team of scientists led by Christopher Polge.[12] This led to a much wider use of cryopreservation today, with many organs, tissues and cells routinely stored at low temperatures. Large organs such as hearts are usually stored and transported, for short times only, at cool but not freezing temperatures for transplantation. Cell suspensions (like blood and semen) and thin tissue sections can sometimes be stored almost indefinitely in liquid nitrogen temperature (cryopreservation). Human sperm, eggs, and embryos are routinely stored in fertility research and treatments. Controlled-rate and slow freezing are well established techniques pioneered in the early 1970s which enabled the first human embryo frozen birth (Zoe Leyland) in 1984. Since then, machines that freeze biological samples using programmable steps, or controlled rates, have been used all over the world for human, animal, and cell biology – 'freezing down' a sample to better preserve it for eventual thawing, before it is deep frozen, or cryopreserved, in liquid nitrogen. Such machines are used for freezing oocytes, skin, blood products, embryo, sperm, stem cells, and general tissue preservation in hospitals, veterinary practices, and research labs. The number of live births from 'slow frozen' embryos is some 300,000 to 400,000 or 20% of the estimated 3 million *in vitro* fertilized births. Dr Christopher Chen, Australia, reported the world's first pregnancy using slow-frozen oocytes from a British controlled-rate freezer in 1986.

Cryosurgery (intended and controlled tissue destruction by ice formation) was carried out by James Arnott in 1845 in an operation on a patient with cancer. Cryosurgery is not common.

Preservation techniques

Cryobiology as an applied science is primarily concerned with low-temperature preservation. Hypothermic storage is typically above 0 °C but below normothermic (32 °C to 37 °C) mammalian temperatures. Storage by cryopreservation, on the other hand, will be in the −80 to −196 °C temperature range. Organs, and tissues are more frequently the objects of hypothermic storage, whereas single cells have been the most common objects cryopreserved.

A rule of thumb in hypothermic storage is that every 10 °C reduction in temperature is accompanied by a 50% decrease in oxygen consumption.[13] Although hibernating animals have adapted mechanisms to avoid metabolic imbalances associated with hypothermia, hypothermic organs, and tissues being maintained for transplantation require special preservation solutions to counter acidosis, depressed sodium pump activity. and increased intracellular calcium. Special organ preservation solutions such as Viaspan (University of Wisconsin solution), HTK, and Celsior have been designed for this purpose.[14] These solutions also contain ingredients to minimize damage by free radicals, prevent edema, compensate for ATP loss, etc.

Cryopreservation of cells is guided by the "two-factor hypothesis" of American cryobiologist Peter Mazur, which states that excessively rapid cooling kills cells by intracellular ice formation and excessively slow cooling kills cells by either electrolyte toxicity or mechanical crushing.[15] During slow cooling, ice forms extracellularly, causing water to osmotically leave cells, thereby dehydrating them. Intracellular ice can be much more damaging than extracellular ice.

For red blood cells, the optimum cooling rate is very rapid (nearly 100 °C per second), whereas for stem cells the optimum cooling rate is very slow (1 °C per minute). Cryoprotectants, such as dimethyl sulfoxide and glycerol, are used to protect cells from freezing. A variety of cell types are protected by 10% dimethyl sulfoxide.[16] Cryobiologists attempt to optimize cryoprotectant concentration (minimizing both ice formation and toxicity) and cooling rate. Cells may be cooled at an optimum rate to a temperature between −30 and −40 °C before being plunged into liquid nitrogen.

Slow cooling methods rely on the fact that cells contain few nucleating agents, but contain naturally occurring vitrifying substances that can prevent ice formation in cells that have been moderately dehydrated. Some cryobiologists are seeking mixtures of cryoprotectants for full vitrification (zero ice formation) in preservation of cells, tissues, and organs. Vitrification methods pose a challenge in the requirement to search for cryoprotectant mixtures that can minimize toxicity.

Cryobiology in humans

Human gametes and two-, four- and eight-cell embryos can survive cryopreservation at −196 °C for 10 years under well-controlled laboratory conditions.[17]

Cryopreservation in humans with regards to infertility involves preservation of embryos, sperm, or oocytes via freezing. Conception, *in vitro*, is attempted when the sperm is thawed and introduced to the 'fresh' eggs, the frozen eggs are thawed and sperm is placed with the eggs and together they are placed back into the uterus or a frozen embryo is introduced to the uterus. Vitrification has its glitches and is not as reliable or proven as freezing fertilized sperm, eggs, or embryos as traditional slow freezing methods because eggs alone are extremely sensitive to temperature. Many researchers are also freezing ovarian tissue in conjunction with the eggs in hopes that the ovarian tissue can be transplanted back into the uterus, stimulating normal ovulation cycles. In 2004, Donnez of Louvain in Belgium reported the first successful ovarian birth from frozen ovarian tissue. In 1997, samples of ovarian cortex were taken from a woman with Hodgkin's lymphoma and cryopreserved in a (Planer, UK) controlled-rate freezer and then stored in liquid nitrogen. Chemotherapy was initiated after the patient had premature ovarian failure. In 2003, after freeze-thawing, orthotopic autotransplantation of ovarian cortical tissue was done by laparoscopy and after five months, reimplantation signs indicated recovery of regular ovulatory cycles. Eleven months after reimplantation, a viable intrauterine pregnancy was confirmed, which resulted in the first such live birth – a girl named Tamara.[18]

Therapeutic hypothermia, e.g. during heart surgery on a "cold" heart (generated by cold perfusion without any ice formation) allows for much longer operations and improves recovery rates for patients.

2.2.4 Scientific societies

The Society for Cryobiology was founded in 1964 to bring together those from the biological, medical, and physical sciences who have a common interest in the effects of low temperatures on biological systems. As of 2007, the Society for Cryobiology had about 280 members from around the world, and one-half of them are US-based. The purpose of the Society is to promote scientific research in low temperature biology, to improve scientific understanding in this field, and to disseminate and apply this knowledge to the benefit of mankind. The Society requires of all its members the highest ethical and scientific standards in the performance of their professional activities. According to the Society's bylaws, membership may be refused to applicants whose conduct is deemed detrimental to the Society; in 1982, the bylaws were amended explicitly to exclude "any practice or application of freezing deceased persons in the anticipation of their reanimation", over the objections of some members who were cryonicists, such as Jerry Leaf.[19] The Society organizes an annual scientific meeting dedicated to all aspects of low-temperature biology. This international meeting offers opportunities for presentation and discussion of the most up-to-date research in cryobiology, as well as reviewing specific aspects through symposia and workshops. Members are also kept informed of news and forthcoming meetings through the Society newsletter, News Notes. The 2011-2012 president of the Society for Cryobiology was John H. Crowe.[20]

The Society for Low Temperature Biology was founded in 1964 and became a registered charity in 2003[21] with the purpose of promoting research into the effects of low temperatures on all types of organisms and their constituent cells, tissues, and organs. As of 2006, the society had around 130 (mostly British and European) members and holds at least one annual general meeting. The program usually includes both a symposium on a topical subject and a session of free communications on any aspect of low-temperature biology. Recent symposia have included long-term stability, preservation of aquatic organisms, cry-

opreservation of embryos and gametes, preservation of plants, low-temperature microscopy, vitrification (glass formation of aqueous systems during cooling), freeze drying and tissue banking. Members are informed through the Society Newsletter, which is presently published three times a year.

2.2.5 Journals

Cryobiology (publisher: Elsevier) is the foremost scientific publication in this area, with about 60 refereed contributions published each year. Articles concern any aspect of low-temperature biology and medicine (e.g. freezing, freeze-drying, hibernation, cold tolerance and adaptation, cryoprotective compounds, medical applications of reduced temperature, cryosurgery, hypothermia, and perfusion of organs).

Cryo Letters is an independent UK-based rapid communication journal which publishes papers on the effects produced by low temperatures on a wide variety of biophysical and biological processes, or studies involving low-temperature techniques in the investigation of biological and ecological topics.

Biopreservation and Biobanking (formerly Cell Preservation Technology) is a peer-reviewed quarterly scientific journal published by Mary Ann Liebert, Inc. dedicated to the diverse spectrum of preservation technologies including cryopreservation, dry-state (anhydrobiosis), and glassy-state and hypothermic maintenance. Cell Preservation Technology has been renamed Biopreservation and Biobanking and is the official journal of International Society for Biological and Environmental Repositories.

2.2.6 See also

- Cryptobiosis
- Insect winter ecology

2.2.7 References

[1] Maki LR, Galyan EL, Chang-Chien MM, Caldwell DR (1974). "Ice Nucleation Induced by Pseudomonas syringae". Applied Microbiology 28 (3): 456–459. PMC 186742. PMID 4371331.

[2] Zachariassen KE, Kristiansen E (2000). "Ice nucleation and antinucleation in nature". Cryobiology 41 (4): 257–279. doi:10.1006/cryo.2000.2289. PMID 11222024.

[3] Food Safety Watch

[4] Scientist finds fungus gnats survive winter half-frozen | Alaskana | ADN.com

[5] Alaska beetles survive 'unearthly' temperatures

[6] ADW: Lithobates sylvaticus: INFORMATION

[7] Nation & World | Looking to frozen frogs for clues to improve human medicine | Seattle Times Newspaper

[8] http://virgil.ruc.dk/~{}cryolab/research.html

[9] ftvert

[10] Cryobiology

[11] Barnes, Brian M. (30 June 1989). "Freeze Avoidance in a Mammal: Body Temperatures Below 0°C in an Arctic Hibernator" (PDF). *Science* (American Association for the Advancement of Science) **244** (4912): 1521–1616. doi:10.1126/science.2740905. PMID 2740905. Retrieved 2008-11-23.

[12] C. Polge, AU Smith, AS Parkes, Revival of spermatozoa Effective vitrification and dehydration at low temperatures Nature, 164 (1949), p. 666

[13] Raison JK (1973). "The influence of temperature-induced phase changes on the kinetics of respiratory and other membrane-associated enzyme systems". *J Bioenerg* **4** (1): 285–309. doi:10.1007/BF01516063. PMID 4577759.

[14] Mühlbacher F, Langer F, Mittermayer C (1999). "Preservation solutions for transplantation". *Transplant Proc* **31** (5): 2069–70. doi:10.1016/S0041-1345(99)00265-1. PMID 10455972.

[15] Mazur P (1977). "The role of intracellular freezing in the death of cells cooled at supraoptimal rates". *Cryobiology* **14** (3): 251–72. doi:10.1016/0011-2240(77)90175-4. PMID 330113.

[16] Hunt CJ, Armitage SE, Pegg DE (2003). "Cryopreservation of umbilical cord blood: 1. Osmotically inactive volume, hydraulic conductivity and permeability of CD34(+) cells to dimethyl sulphoxide". *Cryobiology* **46** (1): 61–75. doi:10.1016/S0011-2240(02)00180-3. PMID 12623029.

[17] "Freezing". Pacific Fertility Center. 2010. Retrieved 2010-02-28.

[18] "Livebirth after orthotopic transplantation of cryopreserved ovarian tissue" (PDF). J Donnez, M M Dolmans, D Demylle, P Jadoul, C Pirard, J Squifflet, B Martinez-Madrid, A Van Langendonckt. 2004. Retrieved 2015-01-02.

[19] Darwin, Mike (1991). "Cold War: The Conflict Between Cryonicists and Cryobiologists". *Cryonics* (Alcor Life Extension Foundation). Retrieved 2009-08-24.

[20] "Officers & Governors". Society for Cryobiology. Retrieved 2010-03-13.

[21] (Charity Commission for England & Wales No. 1099747)

2.2.8 External links

- Society for Cryobiology
- Society for Low Temperature Biology
- CRYOBIOLOGY
- CryoLetters
- Cell Preservation Technology
- Cellular cryobiology and anhydrobiology
- An overview of the science behind cryobiology at the Science Creative Quarterly

2.3 Cryonics Institute

The **Cryonics Institute** (**CI**) is a member-owned-and-operated not-for-profit corporation which provides cryonics services, regarding the preservation of humans in liquid nitrogen after legal death, with hopes of restoring them when new technology will be developed in the future. The Cryonics Institute continues to be an industry leader in terms of both membership & practical affordability for all and also the Cryonics Organization with the largest number of members worldwide (Fully & Not Funded). CI is located in Clinton Township, Michigan.

As of September 1st, 2015, The Cryonics Institute has 1,349 members in total (including preserved bodies). 185 of those funded members had contracts with Suspended Animation, Inc. for standby & transport. 135 humans, 190 human tissue/DNA samples and 112 pets and 60 pet tissue/DNA samples are cryogenically preserved in liquid nitrogen storage at the Cryonics Institute's Michigan facility.[3]

2.3.1 History

The Cryonics Institute was incorporated in Michigan on 4 April 1976 by four local residents: Richard C. Davis, Robert Ettinger, Mae A. Junod and Walter E. Runkel. CI's first client was Ettinger's mother in 1977, and until the beginning of the 1990s, the only other client was Ettinger's first wife in 1987.[4]

In March 1978, The Cryonics Institute purchased a building near Detroit.[4] It served as its location until 1994, when the organization moved to the new *Erfurt Runkel Building*. It is named after *John C Erfurt* and *Walter E. Runkel* (who are now both in suspension there), and has a sprinkler system for additional security.[5][6]

Robert Ettinger was CI President for over 25 years until September 2003, when Ben Best became President/CEO and Robert Ettinger became Vice-President. Mr Ettinger retired as Vice-President on his 87th birthday in December 2005, but remained a Director until new directors were elected in September 2006. For most of the 1990s, Benjamin Best was President of the Cryonics Society of Canada (CSC) and was Editor of *Canadian Cryonics News* until the last issue was published in Spring of 2000. He is still a Director of CSC.

In 2003, an article was published in *Sports Illustrated* magazine centering around the cryonics organization Alcor; the article contained accusations from a fired Alcor employee alleging Alcor had mishandled the cryopreservation of baseball player Ted Williams. Although the Cryonics Institute was not involved in the case, the media hype spurred the state of Michigan to place CI under a "Cease and Desist" order for six months.

Subsequently, the Michigan government decided to license and regulate the Cryonics Institute as a cemetery.[7] As a result, the perfusion of the bodies could not be performed in the Cryonics Institute building from 2004 to 2012. In accordance to law, this was done at the facilities of a funeral director. In 2012, the new Michigan Republican government reversed the Cryonics Institute's classification as a cemetery, removing it from cemetery regulation.[8]

2.3.2   Organization

Cryonics Institute main facility in Clinton Township, Michigan

The Cryonics Institute has 12 Directors on its Board of Directors[9] four of whom are elected by the members every year at the Annual General Meeting (usually held on the last Sunday of September). The Board then selects the Officers: President, Vice-President, Secretary and Treasurer. All members of the board are volunteers.[10]

All Officers[11] of the Cryonics Institute are also Directors.[12] As of September, 2014, the Cryonics Institute Directors & Officers were:

2.3.3   Policies

The Cryonics Institute only allows its members to arrange for whole body storage, not heads (neuropreservation).

The basic $28,000/$35,000 cryopreservation fees and contract with the Cryonics Institute does not include Standby or Transport. CI members living outside of Michigan must normally provide extra funding (Less than $5000) to pay for Funeral Director services and shipping. CI members in the United States wanting Standby & Transport from cryonics professionals can contract for additional payment to the Florida-based company Suspended Animation, Inc.[13] Cryonics Institute members in Canada wanting Preparation (Perfusion CI-VM-1) & Transport can contract for additional payment between 3,000$CDN & 5,000$CDN with Magnus Poirier Team 24/24, 7 days a week. CI has had clients from as far away as India or New Zealand.

2.3.4   Technical procedures

The Cryonics Institute has always provided initial procedures, transport and storage internally, without contracting out to other providers.[14] For most of its history, CI perfused bodies with the (antifreeze) cryoprotectant glycerol, but in the year 2000 a cryobiologist was hired: Yuri Pichugin, Ph.D.

CI's Scientific Advisory Board currently consists of Peter Gouras, Henry R. Hirsch, Raphael Haftka, Klaus Sames, Ronald G. Havelock, Yuri Pichugin, Robert Duncan Enzmann & Gunter Boden. CI patients can fund the procedure through life insurance policies which name Cryonics Institute as the beneficiary or pay by pre-payment. Members who have signed up wear medical alert bracelets informing hospitals & Dr's to notify CI in case of any emergency; in the case of a person who is known to be near death in USA (Mainland), CI can send a team for remote standby by/with SA.

2.3.5   Research

At The Cryonics Institute, Pichugin developed a vitrification mixture; the first human client received it in the summer of 2005 using a new procedure in which the head was vitrified while still attached to the body, which was frozen without cryoprotectant.[15] In February 2007 the Cryonics Institute abandoned its efforts to patent its vitrification mixture and disclosed the formula to preclude others from preventing its use by CI.[16] Dr. Pichugin resigned from the Cryonics Institute in December 2007.[17]

In the summer of 2005, the Cryonics Institute obtained custom-built computer-controlled cooling boxes, with LabVIEW software which would allow controlled cooling to a temperature as low as −192 °C (−313 °F). This equipment was necessary for effective application of vitrification, because cooling should be as fast as possible prior to the solidification temperature of the vitrification mixture (about −125 °C), but cooling should be very slow below that temperature to reduce cracking due to thermal stress.

Instead of using dewars for storage, The Cryonics Institute cryopreserves bodies in large fiberglass/resin liquid-nitrogen-filled "thermos bottles" which CI calls "cryostats". The first cryostats were hand-built in-house[18] by Facilities Manager Andy Zawacki, but now the units are custom built by an external manufacturer. Costs for liquid nitrogen in the newest and most efficient cryostats was below $100 per human body per year in May 2006. Cost reduction is greatly assisted by the use of a 3,000 gallon bulk tank for liquid nitrogen, which is located behind the building. From this central point the liquid nitrogen gets distributed to the cryostats over a system of pipes. Liquid nitrogen is refilled on a weekly basis and does not need electricity to operate.

Neuropreservation

Neuropreservation refers to the practice of cryopreserving only the head. The theory is that only the information contained in the brain matters, and that a new body could be cloned or regenerated at some point in the future. The Cryonics Institute does not offer this option, partially because it may cause extra damage to the patient, and partially because the idea of "frozen severed heads" may alienate the public.[19]

2.3.6 References

[1] "Guide to Cryonics Procedures". Cryonics Institute. Retrieved 2015-08-22.

[2] "Cryostats for Cryogenic Storage". Cryonics Institute. Retrieved 2015-08-22.

[3] "Cryonics Institute (CI) Statistics Details". Cryonics Institute. Retrieved 2013-09-17.

[4] Bridge, Steve (1992). "Fifteen Years in Cryonics". *Alcor Indiana newsletter* (Alcor Indiana). Retrieved 2009-08-24.

[5] "Long Life" (PDF). Retrieved 2011-02-12.

[6] Best, Ben. "Cryonics Institute Sprinkler System". Retrieved 2011-02-12.

[7] Piet, Elizabeth (2004-02-17). "Cryonics lab one of three in United States". USA Today. Retrieved 2008-05-30.

[8] Cryonics Institute President's Report, Ben Best (May–June 2012). "LONG LIFE magazine" (PDF). Immortalist Society. Retrieved 2014-12-07.

[9] "Board of Directors". Cyronics Institute. Retrieved July 2015.

[10] "Frozen Bodies Reside In Clinton Township". Channel 4 Detroit. Archived from the original on 2005-03-22. Retrieved 2009-08-25.

[11] "Officers of the Cryonics Institute". Cyronics Institute. Retrieved 2012-10-05.

[12] "Directors of the Cryonics Institute". Cyronics Institute. Retrieved 2012-10-05.

[13] Best, Benjamin (2008). "A History of Cryonics" (PDF). *The Immortalist*. Cryonics Institute. Retrieved 2009-08-24.

[14] Mondragon, Carlos (1994). "Defining the Cryonics Institution". *Cryonics and Life Extension Conference*. Alcor Life Extension Foundation. Retrieved 2009-08-23.

[15] Ben Best. "The Cryonics Institute's 69th Client". Cyronics Institute. Retrieved 2007-03-09.

[16] Best, Benjamin (2007-02-23). "Cryonics Institute Vitrification Formula Disclosure". CryoNet. Retrieved 2007-09-02.

[17] "Dr. Pichugin resigns, Chana de Wolf visits". Cryonics Institute. 2007-12-14. Retrieved 2008-05-01.

[18] Darwin, Mike (1991). "Cold War: The Conflict Between Cryonicists and Cryobiologists". *Cryonics* (Alcor Life Extension Foundation). Retrieved 2009-08-24.

[19] Best, Benjamin (2008). "Revival Assets Seminar" (pdf). *The Immortalist* (Cryonics Institute). Retrieved 2009-08-26.

2.3.7 External links

- Cryonics Institute
- Cryonics Europe
- *Long Life* magazine
- Suspended Animation, Inc.

2.4 Cryonics – Freeze Me

Cryonics – Freeze Me (originally titled **Death in the Deep Freeze**) is a television documentary programme created by ZigZag Production for Five in 2006 for in their *Stranger than Fiction* series. The program's main topic is cryonics and mainly features interviews with Alcor Life Extension Foundation staff or Alcor members. The program is narrated by Michael Lumsden.

Interviews with following people are featured (in order of appearance):

- Tanya Jones, Chief Operating Officer, Alcor

- Michael Riskin, Ph.D, Alcor Board of Directors and Member

- Anita Riskin, Alcor Member

- Dr Arthur W. Rowe, Ph.D, Professor of Forensic Medicine, New York University Medical School

- Terry Katz, Alcor Member

- Aubrey de Grey, Ph.D, Biomedical Gerentologist

- Gregory Fahy, Ph.D Vice President and Chief Scientific Officer, 21st Century Medicine

- Regina Pancake, Head of Alcor Stabilisation Team, South California

- Tilly Nydes

- Robin Nydes, Alcor Member

- Professor Ralph Merkle, Ph.D, Georgia Tech College of Computing

- Dr James R. Baker MD, Director Michigan Nanotechnology Institute

2.4.1   See also

- Gerontology

- Cryobiology

- Life extension

2.4.2   External links

- ZigZag Productions page about the program

2.5   Cryoprotectant

A **cryoprotectant** is a substance used to protect biological tissue from freezing damage (i.e. that due to ice formation). Arctic and Antarctic insects, fish and amphibians create cryoprotectants (antifreeze compounds and antifreeze proteins) in their bodies to minimize freezing damage during cold winter periods. Cryoprotectants are also used to preserve living materials in the study of biology and to preserve food products.

2.5.1   Mechanism

Cryoprotectants operate by increasing the solute concentration in cells. However, in order to be biologically viable they must easily penetrate cells and must not be toxic to cells.

Glass transition temperature

Some cryoprotectants function by lowering the glass transition temperature of a solution or of a material. In this way, the cryoprotectant prevents actual freezing, and the solution maintains some flexibility in a glassy phase. Many cryoprotectants also function by forming hydrogen bonds with biological molecules as water molecules are displaced. Hydrogen bonding in aqueous solutions is important for proper protein and DNA function. Thus, as the cryoprotectant replaces the water molecules, the biological material retains its native physiological structure and function, although they are no longer immersed in an aqueous environment. This preservation strategy is most often utilized in anhydrobiosis.

Toxicity

Mixtures of cryoprotectants have less toxicity and are more effective than single-agent cryoprotectants. A mixture of formamide with DMSO (dimethyl sulfoxide), propylene glycol, and a colloid was for many years, the most effective of all artificially created cryoprotectants. Cryoprotectant mixtures have been used for vitrification (i.e. solidification without crystal ice formation). Vitrification has important applications in preserving embryos, biological tissues, and organs for transplant. Vitrification is also used in cryonics in an effort to eliminate freezing damage.

2.5.2   Conventional cryoprotectants

Conventional cryoprotectants are glycols (alcohols containing at least two hydroxyl groups), such as ethylene glycol , propylene glycol, and glycerol. Ethylene glycol is commonly used as automobile antifreeze, and propylene glycol has been used to reduce ice formation in ice cream. Dimethyl sulfoxide (DMSO) is also regarded as a conventional cryoprotectant. Glycerol and DMSO have been used for decades by cryobiologists to reduce ice formation in sperm,[1] oocytes,[2] and embryos that are cold-preserved in liquid nitrogen.

2.5.3   Examples in nature

Insects

Insects most often use sugars or polyols as cryoprotectants. One species that uses cryoprotectant is *Polistes exclamans*. In this species, the different levels of cryoprotectant can be used to distinguish between morphologies.[3]

Amphibians

Arctic frogs use glucose,[4] but Arctic salamanders create glycerol in their livers for use as a cryoprotectant.

2.5.4 Food preservation

Cryoprotectants are also used to preserve foods. These compounds are typically sugars that are inexpensive and do not pose any toxicity concerns. For example, many (raw) frozen chicken products contain a "solution" of water, sucrose, and sodium phosphates.

2.5.5 Common cryoprotectants

2.5.6 See also

- Antifreeze protein

- Lyophilization

- List of emerging technologies

- Cryopreservation

- Cryostasis (clathrate hydrates)

2.5.7 References

[1] Imrat, P.; Suthanmapinanth, P.; Saikhun, K.; Mahasawangkul, S.; Sostaric, E.; Sombutputorn, P.; Jansittiwate, S.; Thongtip, N.; et al. (February 2013). "Effect of pre-freeze semen quality, extender and cryoprotectant on the post-thaw quality of Asian elephant (Elephas maximus indicus) semen". *Cryobiology* **66** (1): 52–59. doi:10.1016/j.cryobiol.2012.11.003.

[2] Karlsson, Jens O.M.; Szurek, Edyta A.; Higgins, Adam Z.; Lee, Sang R.; Eroglu, Ali (February 2014). "Optimization of cryoprotectant loading into murine and human oocytes". *Cryobiology* **68** (1): 18–28. doi:10.1016/j.cryobiol.2013.11.002.

[3] J.E. Strassmann; R.E. Lee Jr.; R.R. Rojas & J.G Baust (1984). "Caste and sex differencesin cold-hardiness in the social wasps, Polistes annularis and P. exclamans". *Insectes Sociaux* **31** (3): 291–301. doi:10.1007/BF02223613.

[4] Larson, D. J.; Middle, L.; Vu, H.; Zhang, W.; Serianni, A. S.; Duman, J.; Barnes, B. M. (15 April 2014). "Wood frog adaptations to overwintering in Alaska: New limits to freezing tolerance". *Journal of Experimental Biology*. doi:10.1242/jeb.101931.

2.6 Information-theoretic death

The term **Information-theoretic death** relates to physical damage to the brain and the loss of information. It is the destruction of the information within a human brain to such an extent that recovery of the original person is theoretically impossible by any physical means. The concept of information-theoretic death emerged in the 1990s as a response to the progress of medical technology since conditions previously considered as death, such as cardiac arrest, are now reversible, so they can no longer define death.[1] The term alludes to information theory in mathematics.

The term *information-theoretic death* is intended to mean death that is absolutely irreversible by any technology, as distinct from clinical death and legal death, which denote limitations to contextually-available medical care rather than the true theoretical limits of survival. In particular, the prospect of brain repair using molecular nanotechnology raises the possibility that medicine might someday be able to resuscitate patients even hours after the heart stops.

The paper *Molecular Repair of the Brain* by Ralph Merkle defined information-theoretic death as follows:[2]

> A person is dead according to the information-theoretic criterion if their memories, personality, hopes, dreams, etc. have been destroyed in the information-theoretic sense. That is, if the structures in the brain that encode memory and personality have been so disrupted that it is no longer possible in principle to restore them to an appropriate functional state, then the person is dead. If the structures that encode memory and personality are sufficiently intact that inference of the memory and personality are feasible in principle, and therefore restoration to an appropriate functional state is likewise feasible in principle, then the person is not dead.

The exact timing of information-theoretic death is currently unknown. It has been suggested[3] [4] to occur gradually after many hours of clinical death at room temperature as the brain undergoes autolysis. It may also occur more rapidly if there is no blood flow to the brain during life support, leading to the decomposition stage of brain death, or during the progression of degenerative brain diseases that cause extensive loss of brain structure.

The use of information-theoretic criteria has formed the basis of ethical arguments that state that cryonics is an attempt to save lives rather than being an interment method for the dead. In contrast, if cryonics cannot be applied before information-theoretic death occurs, or if the cryopreservation procedure itself causes information-theoretic death, then cryonics is not feasible. Exactly when complete and total information-theoretic death might occur with respect to different types of preservation and decomposition might also be relevant to the speculative field of mind uploading.

Although the idea of information-theoretic death was first introduced in the context of cryonics,[5] the term has since been used in medical journals discussing issues surrounding brain death[6][7][8] with the same meaning first defined by Merkle.

2.6.1 References

[1] IMR (International Medical Rights)

[2] Merkle, Ralph (January–April 1994), "Molecular Repair of the Brain", *Cryonics*, retrieved 2014-12-27 – via Alcor library online

[3] Merkle, R (1992). "The technical feasibility of cryonics". *Medical Hypotheses* (Elsevier) **39** (1): 6–16. doi:10.1016/0306-9877(92)90133-W. PMID 1435395.

[4] Wowk, B (2014). "The future of death". *Journal of Critical Care* (Elsevier) **29** (6): 1111–1113. doi:10.1016/j.jcrc.2014.08.006. PMID 25194588.

[5] Merkle, R (1992). "The technical feasibility of cryonics". *Medical Hypotheses* (Elsevier) **39** (1): 6–16. doi:10.1016/0306-9877(92)90133-W. PMID 1435395.

[6] Whetstine, L; Streat, S; Darwin, M; Crippen, D (2005). "Pro/con ethics debate: When is dead really dead?". *Critical Care* (BioMed Central) **9** (6): 538–542. doi:10.1186/cc3894. PMID 16356234.

[7] Crippen, DW; Whetstine, L (2007). "Ethics review: Dark angels – the problem of death in intensive care". *Critical Care* (BioMed Central) **11** (1): 202. doi:10.1186/cc5138. PMID 17254317.

[8] Wowk, B (2014). "The future of death". *Journal of Critical Care* (Elsevier) **29** (6): 1111–1113. doi:10.1016/j.jcrc.2014.08.006. PMID 25194588.

2.6.2 External links

- Pro/con ethics debate: When is dead really dead?

- Ethics review: Dark angels-- the problem of death in intensive care

- Albert Einstein's brain and information-theoretic death

- Medical Time Travel by Brian Wowk

2.7 James Bedford

For other people named James Bedford, see James Bedford (disambiguation).

James Hiram Bedford (April 20, 1893 – January 12, 1967) was a University of California psychology professor who wrote several books on occupational counseling.[1] He is the first person whose body was cryopreserved after legal death, and who remains preserved at the Alcor Life Extension Foundation.[2][3][4] In the cryonics community, the anniversary of his cryopreservation is celebrated as "Bedford Day".

2.7.1 Cryonic body preservation

In June 1965, Ev Cooper's Life Extension Society (LES) offered the opportunity to preserve one person free of charge, stating that "the Life Extension Society now has primitive facilities for emergency short term freezing and storing our friend the large homeotherm (man). LES offers to freeze free of charge the first person desirous and in need of cryogenic suspension." Bedford took the opportunity and was established as their candidate. Bedford suffered from kidney cancer that had later metastasized into his lungs, a condition that was untreatable at the time.[5] Bedford left $100,000 to cryonics research in his will, but more than this amount was utilized by Bedford's wife and son in court, having to defend his will and his cryopreservation due to arguments created by other relatives.[5]

Bedford's body was frozen a few hours after his death, due to natural causes related to his cancer.[5] His body was preserved by Robert Prehoda (author of the 1969 book *Suspended Animation*), Dr. Dante Brunol (physician and biophysicist) and Robert Nelson (President of the Cryonics Society of California). Nelson then wrote a book about the subject titled, *We Froze the First Man*. Compared to those employed by modern cryonics organizations, the use of cryoprotectants in Bedford's case was primitive. He was injected with Dimethyl sulfoxide, a compound once thought to be useful for long-term cryogenics, so it is unlikely that his brain was protected. Vitrification was not yet possible, further limiting the possibility of Bedford's eventual recovery. In his first suspended animation stages, his body was stored at Edward Hope's Cryo-Care facility in Phoenix, Arizona, for two years, then in 1969 moved to the Galiso facility in California. Bedford was moved from Galiso in 1973 to Trans Time near Berkeley, California, until 1977, before being stored by Bedford's son for many years.[5]

Bedford's body was maintained in liquid nitrogen by his family in southern California until 1982, when it was then moved to Alcor Life Extension Foundation, and has remained in Alcor's care to the present day.[6] In May 1991, his body's condition was evaluated when he was moved to a new storage dewar. The examiners concluded that "it seems likely that his external temperature has remained at relatively low subzero temperatures throughout the storage interval."[7]

2.7.2 Personal life

Bedford married twice. His first wife, Anna Chandler Rice, died in 1917, the same year she and Bedford were married. Bedford married his second wife, Ruby McLagan in 1920. Bedford and McLagan left behind five children: Doris, Donald, Frances, Barbara and Norman. James Bedford enjoyed traveling extensively and photography.[5]

2.7.3 Bibliography

- *Vocational interests of high-school students.* University of California School of education, Division of vocational education. 1930.

- *Youth and the world's work: Vocational adjustment of youth in the modern world.* Society for Occupational Research. 1938.

- *Vocational interests of secondary school students.* Society for Occupational Research, University of California Station. 1938.

- *Occupational exploration: A guide to personal and occupational adjustment.* Society for Occupational Research. 1941.

- *The veteran and his future job: A guide-book for the veteran.* Society for Occupational Research. 1946.

- *Your future job: A guide to personal and occupational orientation of youth.* Society for Occupational Research. 1950.

- *Your future job: A guide to personal and occupational orientation of youth in the atomic age.* Society for Occupational Research. 1956.

2.7.4 References

[1] Cook, Robert Cecil, ed. *Who's Who in American Education*, 1928, p. 63.

[2] Perry, Mike. "A Freezing Before Bedford's". *Physical Immortality 2(2) 7 (2nd Q 2004)*. Depressed metabolism. Retrieved 2010-12-09.

[3] GALEN PRESS Medical Book Extras SOULS ON ICE

[4] "Dr James Hiram Bedford". Find A Grave. July 26, 2008. Retrieved 2008-08-22.

[5] "Dear Dr. Bedford (and those who will care for you after I do)". Cryonics. July 1991. Retrieved 2009-08-23.

[6] Perry, R. Michael (1992). "Suspension Failures: Lessons from the Early Years". *Cryonics*. Alcor Life Extension Foundation. Retrieved 2009-08-22.

[7] "Evaluation of the Condition of Dr. James H. Bedford After 24 Years of Cryonic Suspension". Alcor Life Extension Foundation. August 1991. Retrieved 2009-08-23.

2.7.5 External links

- Groskinsky, Henry. "Edward Hope prepares Bedford's body". Getty Images.

- "Never Say Die". Time Magazine. Feb 3, 1967. When he died... his physician... began to pack the body in ice... They spent eight hours, sending out periodically for more ice...

- Perry, Mike (July 1991). "The First Suspension". *Cryonics*. For the Record (Alcor).

2.8 Neuropreservation

Neuropreservation is a type of cryonics procedure where the brain is preserved with the intention of future resuscitation and regrowth of a healthy body around the brain.[1] Usually the brain is left within the head for physical protection, so the whole head is cryopreserved.[2] A cryonics patient who undergoes neuropreservation is said to be a neuropatient.

The procedure is often done because vitrification of the entire body is not yet available. Vitrification essentially eliminates the mechanical and chemical damage caused by ice formation,[3][4] at the cost of cryoprotectant toxicity and side effects of dehydration of tissue due to the blood-brain barrier.[5] Although not a direct consequence of vitrification itself, storage of the vitrified brain directly in liquid nitrogen raises the further aspect of fractures,[6] which are fewer in number but larger in scale in vitrified tissue than frozen tissue, a consequence of cooling from Tg (-135 °C) to liquid nitrogen's boiling point (-196 °C).

2.8.1 Advantages

Neuropreservation has several advantages over whole body preservation. It costs less; neuropatients are easier to transport in case of legal, social, or physical problems; it is possible to do a better job of perfusing and therefore cryoprotecting the brain when there is no need to consider other tissues, and its smaller volume allows more rapid and less expensive cooling.[2] Aubrey de Grey has theorized that neuropatients will be revived after procedures have been perfected on whole body patients, and therefore have better chances for revival.[7]

2.8.2 History

Neuropreservation was first proposed in 1965 by cryonics co-creator Evan Cooper, proposed again in a speculative scientific paper by gerontologist George M. Martin in 1971, and independently proposed yet again in 1974 by Mike Darwin, and Fred and Linda Chamberlain. The Chamberlains were the founders of the Alcor Life Extension Foundation. In 1976 Fred's father became the first of many neuropreservation patients at Alcor.[8]

Prior to the year 2000, neuropreservation was performed by surgical separation of the body from the head (called cephalic isolation or "neuroseparation") at the end of cryoprotectant perfusion performed on the upper body via the ascending aorta.[2] After that year, Alcor began performing cephalic isolation before cryoprotectant perfusion, in deep hypothermia, and then using the carotid and vetebral arteries directly for perfusion with cryoprotectants.

As of 2014, Alcor, Oregon Cryonics, and KrioRus are the only cryonics organizations that offer neuropreservation. Other organizations, such as the other major provider, the Cryonics Institute, avoid it because they say it is bad for public relations. Alcor claims there are good technical justifications for neuropreservation, and that they will continue to offer it. Approximately three quarters of the cryonics patients stored at Alcor are neuropatients.

2.8.3 References

[1] http://www.alcor.org/Library/html/neuropreservationfaq.html

[2] Bridge, Steve (1995). "The Neuropreservation Option: Head First into the Future". *Cryonics*. Alcor Life Extension Foundation. Retrieved 2009-08-25.

[3] Fahy, Gregory M.; Wowk, Brian (2015). "Principles of Cryopreservation by Vitrification" **1257**. pp. 30–33. doi:10.1007/978-1-4939-2193-5_2. ISSN 1064-3745. Interestingly, the concentrations generated by freezing actu-

ally exceed the concentrations required for the vitrification of even large living systems

[4] Fahy, Gregory M. (2010). "Cryoprotectant toxicity neutralization". *Cryobiology* **60** (3): S45–S53. doi:10.1016/j.cryobiol.2009.05.005. ISSN 0011-2240. In 1977, Fahy [9] and Fahy and Karow [8] pointed out that damage after freezing and thawing in certain cases is actually correlated not with the amount of ice formed but with the concentration of permeating cryoprotectant during freezing and thawing, and that therefore cryoprotectants can exert damaging effects as they are concentrated in the frozen state. Meryman et al. independently reported in the same year that toxic effects of methanol, ethanol, and ammonium acetate in the frozen state are also detectable in thawed erythrocytes [46]. These non-nucleated cells did not show injury attributable to glycerol or Me2SO in the latter experiments, but evidence continued to emerge in support of putatively toxic effects of cryoprotectants during freezing [2,10,12,14,28], including such effects even in glycerolized erythrocytes [47,50,53], and by 1986 the overall evidence had become quite strong [17].

[5] de Wolf, Chana (September 2013). "Cryopreservation of the Brain: An Update" (PDF). *Cryonics*.

[6] http://www.alcor.org/FAQs/faq02.html

[7] Fryer, Jane (2006-07-29). "The Britons dying to get into the human deep freeze". London: Daily Mail. Retrieved 2009-08-25.

[8] Chamberlain, Fred & Linda (July 16, 2006). "FRC Jr.". Lifepact. Retrieved 2008-05-20.

2.8.4 External links

- Neuropreservation FAQ
- The Case for Neuropreservation
- But What Will The Neighbors Think? A Discourse On The History And Rationale Of Neurosuspension

2.9 Raymond and Monique Martinot

Raymond and Monique (née Leroy) Martinot were a French couple whose quest for cryonic preservation came to an end after a freezer malfunction in 2006.[1]

After Madame Martinot died from ovarian cancer in 1984, her widower, a doctor who once taught medicine in Paris, preserved her body in the cellar of their home, sparking a legal battle due to French legal restrictions on the disposal of corpses. Dr. Martinot also spent decades preparing for his own death. He held to the idea that if he were frozen and preserved eventually he could be brought back to life by 2050 with the help of expanded scientific knowledge. In the 1970s he bought a chateau near Saumur in the Loire Valley and began preparing a steel freezer unit in the chateau's freezer for himself.[1]

The legal battle extended to encompass Raymond's corpse after he died following a stroke in 2002, with a court ruling that the bodies must be buried.[2][3] Their bodies were cremated by their son Remy after a malfunction saw their body temperatures rise to −20 °C from −65 °C for a period of several days.[1]

The German electro-minimal Band Welle: Erdball dedicated a song to her, named "Monique Martinot" on their 1995 album *alles ist möglich*.[4]

2.9.1 References

[1] "Freezer failure ends couple's hopes of life after death" by Angelique Chrisafis, theguardian.co.uk; 16 March 2006; accessed 24 March 2014.

[2]

[3] Profile, editions-hache.com (French)

[4] "Welle erdball Monique Martinot lyrics". Retrieved 16 July 2013.

2.10 Robert Prehoda

Robert Wayne Prehoda (July 7, 1931 - June 11, 2009)[1] was an American chemist and futurist. He participated in the first cryonic suspension of a human being, that of James Bedford. He had a wife, Aline.

2.10.1 Works

- What are the effects of current automation trends in the oil industry on management, unions and the employees?, University of Tulsa, 1957

- Technological forecasting methodology, 1966

- Designing the future: the role of technological forecasting, Chilton Book Co., 1967

- Extended youth: the promise of gerontology, Putnam, 1968

- Suspended animation: the research possibility that may allow man to conquer the limiting chains of time, Chilton Book Co., 1969

- Your Next Fifty Years, Penguin Group (USA) Incorporated, 1980 (ISBN 9780441952212)

2.10.2 References

[1] Death of Robert Prehoda

2.10.3 External links

- Interview with Prehoda, Cryonics Reports, Vol. 4, No. 1, January 1969

- "Alcohol as Fuel" (letter to the editor), *Science News*, Vol. 100, No. 6 (Aug. 7, 1971), p. 88

2.11 Shannon Vyff

Shannon Vyff (born June 10, 1975)[1] is an American transhumanist, cryonicist, and author. She is a director of both the Immortality Institute[2] and the Society for Venturism.[3][4]

Vyff's approaches in promoting transhumanism and cryonics were given exposure when her essay 'Confessions of a Proselytizing Immortalist' was published in 2006 as part of the book: *The Scientific Conquest of Death*. Vyff has also been interviewed by the media on subjects pertaining both to her approaches in promoting cryonics to the public at large (especially to children),[5] and her own motivations for becoming a cryonicist.[6]

Vyff published a science fiction adventure novel for children, *21st Century Kids*, in 2007.[7]

She also is known for practicing Caloric Restriction for life extension purposes, and has been featured on multiple news stories regarding the prospect of life extension.[8][9][10]

2.11.1 Publications

- Vyff, Shannon (March 2007). *21st Century Kids*. Warren Publishing, Inc. p. 280. ISBN 1-886057-00-1. – science fiction adventure novel for children aged 9–12. The protagonists undergo cryopreservation and are reanimated after 200 years. The book raises questions about the technological singularity and time travel.

- Immortality Institute, ed. (October 2004). *The Scientific Conquest of Death: Essays on Infinite Lifespans*. Libros en Red. p. 296. ISBN 987-561-135-2. - a collection of essays. Vyff is the author of the chapter, "Confessions of a Proselytizing Immortalist".[11][12]

2.11.2 Personal life

Vyff is a mother to four children, each of whom are signed up to cryonics.[13]

2.11.3 References

[1] Humanity+ Profile

[2] "Immortality Institute Leadership". Immortality Institute. Retrieved 2010-03-27.

[3] Nye, James (January 24, 2013). "Woman, 23, has her head frozen so she can be reborn after a cure for her brain cancer is found - against the wishes of her religious family". *MailOnline*. Daily Mail. Retrieved 2013-01-28.

[4] Dvorsky, George (January 21, 2013). "23-Year Old Kim Suozzi Undergoes Cryonic Preservation After Successful Fundraising Campaign". *Futurism*. Io9. Retrieved 2013-01-28.

[5] "Radio Interview with Dr. James J. Hughes". The Institute for Ethics and Emerging Technologies. Retrieved 2010-03-21.

[6] "Freezing People into the Future: Uncover the Science Behind Cryonics". ABC News. Retrieved 2010-03-27.

[7] Perry, Mike (2007). "21ST Century Kids" (PDF). *Cryonics* (Alcor Life Extension Foundation) **28** (4): 6.

[8] http://abcnews.go.com/2020/story?id=124351

[9] http://abcnews.go.com/Health/Longevity/story?id=4544003

[10] http://www.alcor.org/blog/update_abc_news_barbara_walter/

[11] http://www.imminst.org/book#Vyff

[12] http://www.cbc-network.org/2006/11/book-review-the-scientific-conquest-of-death-

[13] "Freezing People into the Future: Uncover the Science Behind Cryonics". ABC News. Retrieved 2010-03-27.

2.11.4 External links

- Despres, Jonathan (Jan 26, 2008). "Interview with Shannon Vyff".

- Immortality Institute Q&A With Chairperson Shannon Vyff (UKH+) (Nov 29, 2009) on YouTube

- "Living with children while practicing calorie restriction". Evidenced Based Cryonics. July 3, 2008.

2.12 Suspended animation

For other uses, see Suspended animation (disambiguation). "Anabiosis" redirects here. For academic journal of same name, see Journal of Near-Death Studies.

Suspended animation is the slowing or stopping of life processes by exogenous or endogenous means without termination. Breathing, heartbeat, and other involuntary functions may still occur, but they can only be detected by artificial means.

Tiny organisms (e.g. embryos up to eight cells) can be cryogenically preserved and revived. Some have been kept in suspended animation for as long as 13 years.[1][2]

Placing astronauts in suspended animation has been proposed as one way for an individual to reach the end of an interstellar or intergalactic journey, avoiding the necessity for a gigantic generation ship; occasionally the two concepts have been combined, with generations of "caretakers" supervising a large population of frozen passengers.

Since the 1970s, induced hypothermia has been performed for some open-heart surgeries as an alternative to heart-lung machines. Hypothermia, however, provides only a limited amount of time in which to operate and there is a risk of tissue and brain damage for prolonged periods.

2.12.1 Experiments

Temperature-induced

Lowering the temperature of a substance reduces chemical activity by the Arrhenius equation. This includes life processes such as metabolism.

Hypothermic Range In June 2005 scientists at the University of Pittsburgh's Safar Center for Resuscitation Research announced they had managed to place dogs in suspended animation and bring them back to life, most of them without brain damage, by draining the blood out of the dogs' bodies and injecting a low temperature solution into their circulatory systems, which in turn keeps the bodies alive in stasis. After three hours of being clinically dead, the dogs' blood was returned to their circulatory systems, and the animals were revived by delivering an electric shock to their hearts. The heart started pumping the blood around the body, and the dogs were brought back to life.[3]

On 20 January 2006, doctors from the Massachusetts General Hospital in Boston announced they had placed pigs in suspended animation with a similar technique. The pigs were anaesthetized and major blood loss was induced, along with simulated - via scalpel - severe injuries (e.g. a punctured aorta as might happen in a car accident or shooting). After the pigs lost about half their blood the remaining blood was replaced with a chilled saline solution. As the body temperature reached 10 °C (50 °F) the damaged blood vessels were repaired and the blood was returned.[4] The method was tested 200 times with a 90% success rate.[5]

From May 2014, a team of surgeons from UPMC Presbyterian Hospital in Pittsburgh plan to try the above method in gunshot victims (or those suffering from similar traumatic injuries). The trials will be done on ten such severely wounded patients and compared with ten others in similar situation but who had no access to the above method. They currently refer to the procedure as Emergency Preservation and Resuscitation for Cardiac Arrest from trauma.[6]

Cryogenic Range   This concept is speculative as well as frequently misunderstood. Human beings are unable to survive suspended animation at cryogenic (extremely cold) temperatures naturally due to damage from ice formation. The limits of current technology are also insufficient to prevent loss of cellular viability. Cryonics operates under a fundamentally distinct paradigm from suspended animation in that it depends on Future technology as part of its premise for working.

> Suspended animation is distinct from cryonics because it does not require this "benefit of the doubt" concerning future technology. It is something that immediately and demonstrably works. The medical use of suspended animation will still require optimism that diseases can be cured.
> — Brian Wowk[7]

In order to achieve suspended animation, a reliable method to prevent damage to cells would be needed. Vitrification can achieve this in the laboratory only for small amounts of tissue due to cooling and other physical limits combined with cryoprotectant toxicity.[8] There is also only limited evidence that it is possible in principle, because only very small organisms can be vitrified or frozen safely. Research on Caenorhabditis elegans has shown that memories can be recovered, and such organisms can survive vitrification with around 100% success rates.[9]

Chemically induced

An article in the 22 April 2005 issue of the scientific journal *Science* reports success towards inducing suspended animation-like hypothermia in mice. The findings are significant, as mice do not hibernate in nature. The laboratory of Mark B. Roth at the Fred Hutchinson Cancer Research Center in Seattle, Washington, placed the mice in a chamber containing 80 ppm hydrogen sulfide for a duration of 6 hours. The core body temperature of the mice dropped to 13 degrees Celsius and metabolism, as assayed by carbon dioxide production and oxygen use, decreased 10-fold.[10] They also induced hypoxia on nematode embryos and zebrafish embryos, placing them in suspended animation for hours, and then re-animating them simply by returning the oxygen to the embryos.

Massachusetts General Hospital in Boston announced they had been able to hibernate mice using the same method. Their heart rate was slowed down from 500 to 200 beats per minute, respiration fell from 120 to 25 breaths per minute and body temperature dropped to 30 °C (natural: 39 °C). After 2 hours of breathing air without hydrogen sulfide the mice returned to normal. Further studies are needed to see if the gas had damaging effects on the brain, considering the effect of hydrogen sulfide on the body is similar to hydrogen cyanide; it does not slow the metabolic rate but rather inhibits the transfer of energy within the cell via ATP.[11]

Experiments on sedated sheep[12] and partially ventilated anesthetized pigs[13] have been unsuccessful, suggesting that application to large mammals may not be feasible. In any case, long term suspended animation has not been attempted.

2.12.2   Human hibernation

Main article: Therapeutic hypothermia

There are many research projects currently investigating how to achieve "induced hibernation" in humans.[14][15] This ability to hibernate humans would be useful for a number of reasons, such as saving the lives of seriously ill or injured people by temporarily putting them in a state of hibernation until treatment can be given.

Actual and anecdotal cases of suspected human hibernation or states similar to hibernation exist in the literature:

- **Anna Bågenholm**, a Swedish radiologist who survived 40 minutes under ice in a frozen lake in state of cardiac arrest and survived with no brain damage in 1999.

- **Mitsutaka Uchikoshi**, a Japanese man who survived the cold for 24 days in 2006 without food or water when he fell into a state similar to hibernation[16]

- **Paulie Hynek**, who, at age 2, survived several hours of hypothermia-induced cardiac arrest and whose body temperature reached 64 °F (18 °C)[17]

- **John Smith**, a 14-year-old boy who survived 15 minutes under ice in a frozen lake before paramedics arrived to pull him onto dry land and saved him.[18]

2.12.3 See also

- Cryptobiosis

- Life extension

- Suspended animation in fiction

2.12.4 References

[1] "Longest frozen embryo baby born". BBC News. 6 July 2005. Retrieved 14 January 2009.

[2] "Triplets born 13 years apart". Times Online. 6 July 2005. Retrieved 24 January 2010.

[3] Mihm, Stephen (11 December 2005). "Zombie Dogs". *The New York Times*.

[4] "Does the rate of rewarming from profound hypothermi... J Trauma. 2006 - PubMed - NCBI". Ncbi.nlm.nih.gov. 2013-03-25. Retrieved 2013-06-23.

[5] "Doctors claim suspended animation success". The Sydney Morning Herald. 20 January 2006. Retrieved 10 October 2006.

[6] "Left between life and death: First 'suspended animation' trials set to begin in bid to buy time for stabbing and gunshot victims".

[7] Cryonet 2008 Suspended Animation vs Cryonics

[8] Fahy, Gregory M.; Wowk, Brian (2015). "Principles of Cryopreservation by Vitrification" **1257**. pp. 30–33. doi:10.1007/978-1-4939-2193-5_2. ISSN 1064-3745.

[9] Vita-More, Barranco. "Persistence of Long-Term Memory in Vitrified and Revived C. elegans." *Rejuvenation Research* doi: 10.1089/rej.2014.1636

[10] Blackstone, E.; Morrison, M.; Roth, M. (2005). "H2S induces a suspended animation-like state in mice.". *Science* **308** (5721): 518. doi:10.1126/science.1108581. PMID 15845845.

[11] "Gas induces 'suspended animation'". BBC News. 9 October 2006. Retrieved 10 October 2006.

[12] Haouzi P; Notet V; Chenuel B; Chalon B; Sponne I; et al. (2008). "H2S induced hypometabolism in mice is missing in sedated sheep". *Repiratory Physiology &Amp; Neurobiology* **160** (1): 109–115. doi:10.1016/j.resp.2007.09.001. PMID 17980679.

[13] Li, Jia; Zhang, Gencheng; Cai, Sally; Redington, Andrew N (January 2008). "Effect of inhaled hydrogen sulfide on metabolic responses in anesthetized, paralyzed, and mechanically ventilated piglets". *Pediatric Critical Care Medicine* **9** (1): 110–112. doi:10.1097/01.PCC.0000298639.08519.0C. PMID 18477923. Retrieved 23 March 2008. (subscription required (help)). H2S does not appear to have hypometabolic effects in ambiently cooled large mammals and conversely appears to act as a hemodynamic and metabolic stimulant.

[14] New Hibernation Technique might work on humans | LiveScience at www.livescience.com

[15] Race to be first to 'hibernate' human beings - Times Online at www.timesonline.co.uk

[16] Japanese man in mystery survival at BBC News

[17] *Eleva boy's story part of national tour to honor Mayo Clinics 150 years* Mayo Clinic

[18] Suspended Animation? How A Boy Survived 15 Minutes Trapped Under Ice In Frozen Lake at Medical Daily

2.12.5 External links

- Mark Roth: Suspended animation is within our grasp

- Pitt scientists resurrect hope of cheating death

- Suspended Animation: Bringing Back the Dead?

- "Stuck Pig" article at Wired.com

- Hydrogen sulfide does not induce hypometabolism in sheep

- Suspended Animation Technology arrives

- Laboratory of Mark B. Roth at the Fred Hutchinson Cancer Research Center

Chapter 3

Cryonics Organizations

3.1 Alcor Life Extension Foundation

The **Alcor Life Extension Foundation**, most often referred to as **Alcor**, is a Scottsdale, Arizona, USA-based nonprofit organization that researches, advocates for and performs cryonics, the preservation of humans in liquid nitrogen after legal death, with hopes of restoring them to full health when new technology is developed in the future.

As of July 31, 2015, Alcor had 1041 members, 170 associate members and 139 humans in cryopreservation, many as neuropatients (86 of Alcor patients were neuropatients or brain preservation patients as of May 2015).[1][4][5][6] Alcor also cryopreserves the pets of members. As of November 15, 2007, there were 33 pets in suspension.[7][8]

Alcor accepts anatomical donations (cryonics cases) under the Uniform Anatomical Gift Act and Arizona Anatomical Gift Act for research purposes, reinforced by a court case in its favor that affirmed a constitutional right to engage in cryopreservation and donate one's body for the purpose.[9] A form of the Uniform Anatomical Gift Act has been passed in all 50 states.[10]

3.1.1 History

The organization was established as a nonprofit organization by Fred and Linda Chamberlain in California in 1972 as the Alcor Society for Solid State Hypothermia (ALCOR).[11] Alcor was named after a faint star in the Big Dipper.[12][13] The name was changed to Alcor Life Extension Foundation in 1977.[14] The organization was conceived as a rational, technology-oriented cryonics organization that would be managed on a fiscally conservative basis. Alcor advertised in direct mailings and offered seminars in order to attract members and bring attention to the cryonics movement. The first of these seminars attracted 30 people.

On July 16, 1976, Alcor performed its first human cryopreservation on Fred Chamberlain's father.[12] That same year, research in cryonics began with initial funding provided by the Manrise Corporation. At that time, Alcor's office consisted of a mobile surgical unit in a large van. Trans Time, Inc., a cryonics organization in the San Francisco Bay area, provided initial preservation procedures and long-term patient storage[15] until Alcor began doing its own storage in 1982.

In 1977, articles of incorporation were filed in Indianapolis by the Institute for Advanced Biological Studies (IABS) and Soma, Inc. IABS was a nonprofit research startup led by a young cryonics enthusiast named Steve Bridge, while Soma was intended as a for-profit organization to provide cryopreservation and human storage services. Its president, Mike Darwin, subsequently became a president of Alcor. Bridge filled the same position many years later.[10] IABS and Soma relocated to California in 1981.[15] Soma was disbanded, while IABS merged with Alcor in 1982.[10]

In 1978, Cryovita Laboratories was founded by Jerry Leaf, who had been teaching surgery at UCLA. Cryovita was a for-profit organization which provided cryopreservation and transport services for Alcor in the 1980s until Leaf's death, at which time Alcor began providing these services on its own.[16] Leaf and Michael Darwin collaborated to bring the first cryonics patient, Dr. James Bedford, who was preserved in 1967, to Alcor's California facility in 1982.

During this time, Leaf also collaborated with Michael Darwin in a series of hypothermia experiments in which dogs were resuscitated with no measurable neurological deficit after hours in deep hypothermia, just a few degrees above zero Celsius. The blood substitute which was developed for these experiments became the basis for the washout solution used at Alcor. Together, Leaf and Darwin developed a standby-transport model for human cryonics cases with the goal of intervening immediately after cardiac arrest and minimizing ischemic injury. Leaf was cryopreserved by Alcor in 1991; since 1992, Alcor has provided its own cryopreservation as well as patient-storage services. Today, Alcor is the only full-service cryonics organization that performs remote standbys.

Alcor grew slowly in its early years. In 1984, it merged with

This "bigfoot" Dewar is custom-designed to contain four wholebody patients and six neuropatients immersed in liquid nitrogen at −196 degrees Celsius. The Dewar is an insulated container which consumes no electric power. Liquid nitrogen is added periodically to replace the small amount that evaporates.

In 1986, a group of Alcor members formed Symbex, a small investment company which funded a building in Riverside, California, for lease by Alcor. Alcor moved from Fullerton, California, to the new building in Riverside in 1987; Timothy Leary appeared at the grand opening.[19] Alcor cryopreserved a member's companion animal in 1986, and two people in 1987. Three human cases were handled in 1988, including the first whole body patient of Alcor's,[9] and one in 1989. At that time, Alcor owned 20% interest in Symbex, with a goal of 51% ownership.[20] In September 1988, Leary announced that he had signed up with Alcor, becoming the first celebrity to become an Alcor member.[19] Leary later switched to a different cryonics organization, CryoCare, and then changed his mind altogether. Alcor's Vice-President, Director, head of suspension team and chief surgeon, Jerry Leaf, died suddenly of a heart attack in 1991.

By 1990, Alcor had grown to 300 members and outgrown its California headquarters, which was the largest cryonics facility in the world.[20] The organization wanted to remain in Riverside County,[20] but in response to concerns that the California facility was also vulnerable to earthquake risk, the organization purchased a building in Scottsdale, Arizona in 1993 and moved its patients to it in 1994.[5]

Alcor has held seven conferences on life extension technologies, with participants such as Eric Drexler, Ralph Merkle, Ray Kurzweil, Aubrey de Grey, Timothy Leary, Barbara Marx Hubbard, and Michael D. West.[21][22][23]

As of May 31, 2015, Alcor has 1,332 members (including 1,037 living members with full arrangements to be cryopreserved). Also as of 2015, the oldest patient (at time of clinical death) to have undergone cryopreservation procedures at Alcor is Rose Selkovitch, A-2340, who was nearly 102 years old at the time.[24][25] The youngest (also as of 2015) is Matheryn Naovaratpong, A-2789, two years old at the time of her cryopreservation.[26]

Alcor's next conference will take place in October 2015.

3.1.2   Research

In 2001, Alcor adapted cryoprotectant formulas from published scientific literature into a more concentrated formula capable of achieving ice-free preservation (vitrification) of the human brain (neurovitrification). In 2005, the vitrification process was applied to the first whole-body subject (as opposed to brain-only).[27] This resulted in vitrification of the brain and conventional cryopreservation of the rest of the body. Work is continuing towards achieving whole-body vitrification, which is limited by the ability to fully circulate the cryoprotectant throughout the body. The vitrification used since 2000 was switched to what Alcor said

the Cryonics Society of South Florida.[17] Alcor counted only 50 members in 1985, which was the year it cryopreserved its third patient. However, during this time researchers associated with Alcor contributed some of the most important techniques related to cryopreservation, eventually leading to today's method of vitrification.[18]

Increasing growth in membership during this period is partially attributed to the 1986 publication of Eric Drexler's *Engines of Creation*, which debuted the idea of nanotechnology and contained a chapter on cryonics.[12]

was a superior solution in 2005.[27] Canadian businessman, Robert Miller, founder of Future Electronics, has provided research funding to Alcor in the past.[28]

3.1.3 Policies and procedures

Alcor is governed by a self-perpetuating board of directors.[29] Alcor's Scientific Advisory Board currently consists of Antonei Csoka, Aubrey de Grey, Robert Freitas, Bart Kosko, James B. Lewis, Ralph Merkle, Marvin Minsky, Martine Rothblatt, and Michael D. West.[30] Alcor also maintains a medical advisory board consisting of medical doctors.[14]

Most Alcor patients fund the procedure through life insurance policies which name Alcor as the beneficiary.[5] Members who have signed up wear medical alert bracelets informing hospitals and doctors to notify Alcor in case of any emergency; in the case of a person who is known to be near death, Alcor can send a team for remote standby.

In some states, members can sign certificates stating that they wish to decline an autopsy. The cutting of the body organs (especially the brain) and blood vessels required for an autopsy makes it difficult to either preserve the body, especially the brain, without damage or perfuse the body with glycerol.[10] The optimum preservation procedure begins less than one hour after death.[10] Members can specify whether they wish Alcor to attempt to preserve even if an autopsy occurs, or whether they wish to be buried or cremated if an autopsy renders little hope for preservation.[10]

In cases with remote standby, cardiopulmonary support is begun as soon as a patient is declared legally dead.[31] Some patients were not able to receive cardiopulmonary support immediately, but in deference to the possibilities of future technology, these patients have also been preserved with the best techniques available.[32] Alcor has a network of paramedics nationwide and seven surgeons, located in different regions, who are on call 24 hours a day.[33] If an Alcor patient is met by a standby team (usually at a hospital, hospice, or home), the team will perform CPR to maintain blood flow to the brain and organs while simultaneously pumping an organ preservation solution through the veins.[34]

Patients are transported as quickly as possible to Alcor headquarters in Scottsdale, where they undergo final preparations in Alcor's cardiopulmonary bypass lab.[35] Plans are underway for a second operating room to be built.[35] In the Patient Care Bay,[36] patients are monitored by computer sensors while kept in liquid nitrogen in dewars.[10] Liquid nitrogen is refilled on a weekly basis and does not need electricity to operate.[36][37] Riverside County, California deputy coroner Dan Cupido said that Alcor had better equipment than some medical facilities.[38]

Membership dues cover one-third of Alcor's yearly budget, with donations and case income from cryopreservations covering the rest.[39] Alcor receives $50,000 each year from television royalties donated by a sitcom writer and producer who are in suspension.[37] In 1997, after a substantial effort led by then-president Steve Bridge, Alcor formed the Patient Care Trust as an entirely separate entity to manage and protect the funding for cryopatients, including owning the building.[37] Alcor remains the only cryonics organization to segregate and protect patient funding in this way; the 2% annual growth of the Trust is enough for upkeep of the patients.[37] At least $115,000 of the money received for each full-body patient goes into this trust for future patient care, $25,000 for a neuropatient.[36] Alcor is currently working to create an Alcor Model Trust, which would make it easier for members to establish their own Trusts to preserve their assets following legal death and prior to being revived from cryopreservation.[40] Some members have already taken steps to do this on their own.[41] Members can also store possessions deep underground in a Kansas salt mine operated by Underground Vaults & Storage, Inc.[42]

Further information about Alcor policies and procedures is available from their FAQs.[43]

3.1.4 Membership

Members suspended include Dick Clair, an Emmy Award-winning television sitcom writer and producer, Hall of Fame baseball legend Ted Williams and his son John Henry Williams, and futurist FM-2030.[13][14]

Notable current members include:[18][40][40][44][45][46][47] researcher Aubrey de Grey, nanotechnology pioneer Eric Drexler, engineer Keith Henson and his family, entrepreneur Saul Kent, inventor Ray Kurzweil,[48] casino owner Don Laughlin,[49][50] film director Charles Matthau, PayPal founder and venture capitalist Peter Thiel,[51] Internet pioneer Ralph Merkle, Canadian businessman Robert Miller,[52] MIT professor Marvin Minsky, futurists Max More and Natasha Vita-More, entrepreneur Luke Nosek, mathematician Edward O. Thorp, talk radio host Mark Edge, and computer security CEO Kenneth Weiss.

Magazine publisher Althea Flynt was signed up to Alcor, but her body was not able to be preserved after her death, which resulted in an autopsy.[53] One Alcor member died in the World Trade Center in the September 11 attacks.[54]

Membership has grown at a rate of about eight percent a year since Alcor's inception,[37] tripling between 1987 and 1990.[55] The oldest patient at Alcor is a 101-year-old woman, and the youngest is a 2-year-old girl.[39][56] Alcor has had patients from as far as Australia.[57] One in four of its members resides in the San Francisco Bay Area.[45]

The membership receives Alcor's magazine, *Cryonics*, published 12 times a year, but it's also available online for free.

3.1.5 Controversies

Dora Kent

Before the company moved to Arizona from Riverside, California in 1994, it became a center of controversy when a county coroner ruled that Alcor client Dora Kent (Alcor board member Saul Kent's mother) was murdered with barbiturates before her head was removed for neuropreservation by the company's staff. Alcor contended that the drug was administered after her death. No charges were ever filed;[58] former Riverside County deputy coroner Alan Kunzman later claimed that this was due to mistakes and poor decision-making by others in his office.[59]

A judge ruled that Kent was already deceased at the time of preservation, and no foul play was involved.[59][60] Alcor sued the county for false arrest and illegal seizure and won both suits.[12] The incident is credited with spurring a growth in membership for Alcor due to the resultant publicity.[12]

Ted Williams

In 2002, Alcor drew considerable attention when baseball star Ted Williams was placed in cryonic suspension; although Alcor maintains privacy of its patients if they wish and did not disclose that Williams was at the Scottsdale facility, the situation came to light in court documents that grew out of an extended family dispute over Williams' wishes in regard to his remains.[61] While Williams' children Claudia and John Henry contended that Williams wished to be preserved at Alcor, their half-sister and oldest Williams child Bobby-Jo Ferrell contested that her father wished to be cremated.[61] Williams' attorney produced a note signed by Williams, John Henry, and Claudia saying: "JHW, Claudia and Dad all agree to be put into biostasis after we die. This is what we want, to be able to be together in the future, even if it is only a chance."[62] John Henry later said, "He was very into science and believed in new technology and human advancement and was a pioneer. Even though things seemed impossible at times, he always knew there was always a chance to catch a fish -- only if you had your fly in the water."[33]

In 2003, *Sports Illustrated* published allegations by former Alcor COO Larry Johnson that the company had mishandled Williams' head by drilling holes and accidentally cracking it. Johnson also claimed that some of Williams'

DNA was missing; the article alleges that Williams' son, John Henry Williams, desired to sell some of his father's DNA, a charge John Henry denied. Williams' attorney called the DNA allegations an "absurd proposition" and accused Johnson of trying to grab headlines.[63] Alcor denied the allegations of missing DNA[64] and explained that microscopic cracking can result as part of the process of freezing the head, damage which is less than previous methods using glycerol during cryopreservation; Alcor believes that technology sufficient to revive its patients would also be able to repair the microscopic fractures, which are monitored using a tiny microphone.[65] In the wake of the *Sports Illustrated* story, Johnson began a paid-membership website where he displayed what he said were photographs of Williams.[66]

John Henry Williams subsequently died of leukemia, and his remains are also stored at Alcor.[67] After John Henry's death, Ferrell again filed a lawsuit, but representatives of Williams' estate repeated that he wished to be at Alcor.[62]

1992 death

In addition to his Williams allegations, Johnson handed over to the police a taped conversation in which he claims Alcor facilities engineer Hugh Hixon stated that an Alcor employee deliberately hastened the imminent 1992 death of a terminally ill AIDS patient, with an injection of Metubine, a paralytic drug.[64] The nurse who pronounced the 1992 death has denied Johnson's claim that there was any hastening of death.[68] The nurse's claim that the patient died in his bedroom contradicts Alcor's own 1992 case report, in which they state the patient died approximately 30 minutes after they transported him to a makeshift operating room, in a garage.[69] In 2009, Carlos Mondragon, (Alcor's CEO at the time of the incident), told ABC News he had been made aware of the allegations, at the time of the case, and as a result, had severed Alcor's ties with the employee who allegedly hastened the patient's death.[70] Mr. Mondragon failed to inform ABC News that the same person later performed Alcor's surgical procedures, including the suspension of Ted Williams.

3.1.6 See also

- Information-theoretic death

3.1.7 References

[1] "Membership and Applicant Growth". Alcor.org. Retrieved 2015-08-19.

[2] "Bylaws of the Alcor Life Extension Foundation". Alcor.org. Retrieved 2013-04-04.

[3] "The Alcor Patient Care Trust". Alcor.org. Retrieved 2013-04-04.

[4] "Alcor Cases". Alcor.org. Retrieved 2013-11-06.

[5] Quigley, Christine (1998). *Modern Mummies: The Preservation of the Human Body in the Twentieth Century.* McFarland. p. 143. ISBN 0-7864-0492-2.

[6] "Membership Numbers Update, May 31, 2015". Alcor.org. 2015-06-08. Retrieved 2015-08-19.

[7] "FAQ - Membership". Alcor. 2011-12-21. Retrieved 2013-04-04.

[8] "Cryopreservation and Related Cases » Alcor News - News blog of the Alcor Life Extension Foundation". Alcor.org. 1999-02-22. Retrieved 2013-11-06.

[9] Mondragon, Carlos (1990). "A Stunning Legal Victory for Alcor". *Cryonics.* Alcor Life Extension Foundation. Retrieved 2009-08-23.

[10] April 26, 1996. Dying to know *ASU Cronkite School of Journalism.*

[11] "Alcor IRS letter" (PDF). *IRS.* Alcor Life Extension Foundation. 1972. Retrieved 2009-08-23.

[12] Best, Ben (2008). "A History of Cryonics" (PDF). *The Immortalist.* Cryonics Institute. Retrieved 2009-08-24.

[13] Kunen, James S. (July 17, 1989). "Reruns Will Keep Sitcom Writer Dick Clair on Ice—indefinitely". People Magazine. Retrieved 2009-08-24.

[14] "Alcor: The Origin of Our Name" (PDF). Alcor Life Extension Foundation. Winter 2000. Retrieved 2009-08-25.

[15] 1981. IABS Suspension Coverage *Cryonics.*

[16] Mondragon, Carlos (1994). "Defining the Cryonics Institution". *Cryonics and Life Extension Conference.* Alcor Life Extension Foundation. Retrieved 2009-08-23.

[17] "Unity and Disunity in Cryonics". *Cryonics.* 1992.

[18] "A Brain Is A Terrible Thing To Waste". *Mensa International.* Retrieved 2009-08-23.

[19] Darwin, Mike (September 1988). "Dr. Leary Joins Up...". Alcor Life Extension Foundation. Retrieved 2009-08-24.

[20] Mondragon, Carlos. 1990. Alcor Begins Planning a New Facility *Cryonics.*

[21] 2007. 7th Alcor Conference Alcor.

[22] November 8, 2002. Eric Drexler, Michael D. West Among Top Speakers for Alcor Life Extension Foundation's 5th Annual Conference. *PR Newswire.*

[23] "Ad for 1978 Alcor conference, featuring many speakers now dead". *Alcor Life Extension Foundation.* Imminst.org forum. 1978. Retrieved 2009-08-24.

[24] "Alcor Cryopreserves 80th Patient - Alcor NewsAlcor News - News Blog of the Alcor Life Extension Foundation". Alcor.org. 2008-03-29. Retrieved 2015-09-16.

[25] "Mariette Selkovitch becomes Alcor's 136th patient on May 5, 2015 - Alcor NewsAlcor News - News Blog of the Alcor Life Extension Foundation". Alcor.org. 2015-05-05. Retrieved 2015-09-16.

[26] "Two-Year Old Thai Girl Becomes Alcor's 134th Patient - Alcor NewsAlcor News - News Blog of the Alcor Life Extension Foundation". Alcor.org. 2015-01-08. Retrieved 2015-09-16.

[27] "New Cryopreservation Technology". Alcor Life Extension Foundation. October 2005. Retrieved 2009-08-25.

[28] "The World's Billionaires: #314 Robert Miller". Forbes. October 2005. Archived from the original on 12 March 2007. Retrieved 2007-03-08.

[29] Merkle, Ralph C. (2008). "Alcor's Self Perpetuating Board". Alcor Life Extension Foundation. Retrieved 2009-08-25.

[30] "Alcor Scientific Advisory Board". Alcor Life Extension Foundation. Retrieved 2009-08-26.

[31] Wowk, Brian. "Cardiopulmonary Support in Cryonics". Alcor Life Extension Foundation. Retrieved 2009-08-25.

[32] "Cases Without Cardiopulmonary Support". Alcor Life Extension Foundation. Retrieved 2009-08-25.

[33] Moehringer, J.R. January 22, 2003. Comeback Would Top Them All *Los Angeles Times.*

[34] August 23, 2005. The Cold, Hard Facts on Cryonics *The Chicago Tribune.*

[35] Jones, Tanya (2008). "Update on Recent Progress". *Alcor News blog.* Alcor Life Extension Foundation. Retrieved 2009-08-25.

[36] Bridge, Steve (1995). "The Neuropreservation Option: Head First into the Future". *Cryonics.* Alcor Life Extension Foundation. Retrieved 2009-08-25.

[37] Sullivan, Will (2006). "Second place non-fiction: 'Iceman Cometh'". Yale Daily News. Retrieved 2009-08-27.

[38] Regis, Ed (1991). *Great Mambo Chicken And The Transhuman Condition: Science Slightly Over The Edge.* Westview Press. p. 103. ISBN 0-201-56751-2.

[39] Vascellaro, Charlie (2007). "Waiting to Awake". *AZ Business* (AZ Big Media). Retrieved 2009-08-26.

[40] Best, Ben (2008). "Asset Preservation" (PDF). *The Immortalist* (Cryonics Institute). Retrieved 2009-08-25.

[41] Regalado, Antonio (2006). "A Cold Calculus Leads Cryonauts To Put Assets on Ice". The Wall Street Journal. Retrieved 2009-08-26.

[42] "Is it possible to place some of my personal effects into storage with Alcor?". Alcor.

[43] "Alcor FAQs". Alcor.

[44] Fryer, Jane (2006-07-29). "The Britons dying to get into the human deep freeze". London: Daily Mail. Retrieved 2009-08-25.

[45] Guynn, Jessica (2002). "Techies go for ice-cold afterlife". *Contra Costa Times*. Retrieved 2009-08-26.

[46] "Links". Max More. Retrieved 2009-08-26.

[47] Sandomir, Richard (2005-02-13). "Please Don't Call the Customers Dead". The New York Times. Retrieved 2009-08-27.

[48] Philipkoski, Kristen (2002-11-18). "Ray Kurzweil's Plan: Never Die". Wired. Retrieved 2009-08-26.

[49] "5 billionaires who want to live forever: Don Laughlin". April 4, 2013. Retrieved 2013-04-04.

[50] "Laughlin Training". *Alcor News blog*. Alcor Life Extension Foundation. 2009. Retrieved 2009-08-25.

[51] "Peter Thiel: the billionaire tech entrepreneur on a mission to cheat death". September 19, 2014. Retrieved 2015-06-08.

[52] "5 billionaires who want to live forever: Robert Miller". April 4, 2013. Retrieved 2013-04-04.

[53] Quigley, Christine (1998). *Modern Mummies: The Preservation of the Human Body in the Twentieth Century*. McFarland. p. 145. ISBN 0-7864-0492-2.

[54] "Splendid Splinter chilling in Scottsdale". Sports Illustrated. Associated Press. 2003. Retrieved 2013-04-04. Those waiting for death pay annual dues — $398 for the first family member. Those who die in accidents probably won't be in good enough shape to be preserved. One member was lost in the World Trade Center disaster.

[55] Cieply, Michael (1990-09-09). "They Freeze Death if Not Taxes - Cryonics: Freezing bodies for later, uh, reanimation gains popularity. But resurrection is not without complications.". Los Angeles Times. Retrieved 2009-08-26.

[56] Jones, Tanya (2008). "Alcor Cryopreserves 80th Patient". *Alcor News* (Alcor Life Extension Foundation). Retrieved 2009-08-26.

[57] Bersten, Rosanne (2002-08-24). "Australians put hands up for big freeze". Melbourne: The Age. Retrieved 2009-08-25.

[58] Perry, Michael (September–November 1992). "OUR FINEST HOURS: Notes On the Dora Kent Crisis". *Cryonics*. Retrieved 2013-04-04.

[59] Fisher, Michael. 2004. Ex-coroner says errors hurt probe. *The Press-Enterprise*.

[60] Perry, R. Michael (2000). *Forever for All: Moral Philosophy, Cryonics, and the Scientific Prospects for Immortality*. Universal-Publishers. p. 40. ISBN 1-58112-724-3.

[61] Associated Press. August 2, 2003. Splendid Splinter chilling in Scottsdale. *Sports Illustrated/CNN*.

[62] Sandomir, Richard. June 17, 2004. Ted Williams Legal Fight Comes to an End *The New York Times*.

[63] Associated Press. July 8, 2002. Dispute over Ted Williams' body divides son, daughter Half brother accused of plan to cryogenically freeze body. *The Seattle Times*.

[64] Bertolino, Bill. 2003. Scottsdale company's role in death probed. *East Valley Tribune*.

[65] Platt, Charles. August 13, 2003. Renewed Ted Williams Controversy: An Interim Response *Alcor News*.

[66] August 13, 2003. The Story of Ted Williams *CNN*.

[67] "ESPN - Leukemia claims son of Hall of Famer - MLB". ESPN.go.com. 2004-03-13. Retrieved 2013-04-04.

[68] Hennes, Ronald (November 11, 2009). "I WAS THE NURSE/I WAS THERE". Amazon.com. Retrieved 2009-12-12.

[69] "Alcor Case Report: Neurosuspension of Patient A-1260". Alcor.org. 2011-12-21. Retrieved 2013-04-04.

[70] Page 2 of 4 (2009-10-07). "Page 2: Former Alcor Employee Makes Harsh Allegations Against Cryonics Foundation - ABC News". Abcnews.go.com. Retrieved 2013-04-04.

3.1.8   External links

- Official website

- Official websites for Mission Statement, *Cryonics* magazine, News Blog, FAQs

- Oberhaus, Daniel (Nov 28, 2014). "The Art of Not Dying". Vice.

3.2   American Cryonics Society

The **American Cryonics Society** (**ACS**), also known as the **Cryonics Society of America**, is a member-run, California-based, 501(c)(3) tax-exempt nonprofit organization that supports and promotes research and education into cryonics and cryobiology. Cryonics is the preservation through cold storage, usually with liquid nitrogen, of humans (and sometimes non-human animals) after legal death. This procedure is done in the hopes of eventual "reanimation." Any such reanimation depends upon future technological advances that are hoped for, but by no means assured or promised.

The American Cryonics Society is the oldest cryonics organization still in existence.[2] Since 1972 ACS has offered a program where members who enroll, are placed into cryonic suspension upon their deaths and then maintained in liquid nitrogen. This program provides for continuous funding so that the relatives of the subject are not required to pay for the initial freezing, yearly maintenance in liquid nitrogen, or eventual reanimation (should the latter prove possible). Members often provide such funding through the purchase of a life insurance policy.

3.2.1 History

The American Cryonics Society was first incorporated in 1969 in San Francisco as the Bay Area Cryonics Society (BACS);[3] its name was changed to the American Cryonics Society in 1985. The founding of the company followed over two years of organizational meetings by cryonics activists. Signers of the founding charter included two well-known Bay Area physicians, Dr. M. Coleman Harris, and Dr. Grace Talbot. The 1969 incorporation date makes it the oldest cryonics society still in existence. The Immortalist Society (IS), with which the American Cryonics Society works closely, is a successor to the Cryonics Society of Michigan whose founding predates that of the American Cryonics Society.

Since its beginning, the American Cryonics Society has made valuable contributions to research and methodology of freezing and cold storage of organs and organism. The first suspensions of humans under the ACS program were in 1974 through Trans Time, a company founded in 1972 by activist members of the American Cryonics Society (then BACS). This was followed by a succession of additional suspensions and ongoing research into methods of preservation and procedures for maintaining tissue, organs, and organisms at liquid nitrogen temperature.

Starting in 1974 ACS-sponsored research into establishing suspension procedures included development of "blood substitutes" and flushes to replace the blood in cryonic patients with a solution which had cryoprotective properties. Dr. Paul Segall was the chief researcher, and following a technological report that members of the press viewed as significant, Dr. Segall, Dr. Richard Marsh, and others appeared on Good Morning America, The Phil Donahue Show, The Sally Jessy Raphael Show, and made many other popular media appearances. Somewhat later, ACS President Dr. Avi Ben-Abraham also did a media blitz to publicize cryonics research. American Cryonics Society researchers would later develop commercial organ preservation solutions, based in part on ACS-sponsored research.

In May 1988, ACS researchers authored a paper, "Interventive Gerentology, Cloning, and Cryonics: Relevance to Life Extension," that was published in *Biomedical Advances* and edited by Allan L. Goldstein. The inclusion of this paper in a publication of serious academic inquiry properly identifies cryonics as scientific discipline.

Concurrent with biological research, the American Cryonics Society established financial and managerial policy to better safeguard the funds of members in suspension. This included funding through life insurance and conservative estimates of suspension and maintenance costs to ensure adequate funding. In 1981, the American Cryonics Society employed attorney Jim Bianchi to develop model trust and model will documents for people who wished suspension. Mr. Bianchi also researched the legal basis for cryonics, and developed a set of related documents.

In 1978, the Cryonics Society of America researchers collaborated with Jerry Leaf of Cryovita Laboratories in experiments that would apply the methodology to cryonic suspensions that were then in use as surgical procedures to treat patients with heart disease. Thus a team led by a thoracic surgeon and a perfusionist would use cardio-pulmonary resuscitation equipment and blood-pumps to quickly cool a patient, and replace his blood with a blood substitute containing cryoprotectants. For several years thereafter, Cryovita Laboratories led by Jerry Leaf and Mike Darwin, were responsible for the initial cryonic suspension of patients of the American Cryonics Society, which were then perfused, transported, and kept in cryogenic storage by Trans Time.[3] This method of making use of contract companies as a means of risk management was unique to the American Cryonics Society and has been followed to the present day.

In 1992, the American Cryonics Society signed contracts with the newly formed CryoSpan Corporation and transferred a number of patients to that Southern California facility, as well as making use of CryoSpan services for a number of new patients. At about this same time it contracted with BioPreservation Inc., operated by Mike Darwin, to perform the standby and initial suspension of the Cryonics Society of America members. The American Cryonics Society also purchased suspension and "first response" equipment from Darwin and other suppliers to enable the American Cryonics Society to freeze its own members, to supplement its employment of contract companies.

In 2002 when CryoSpan opted to close down its long-term cryogenic storage operations, 10 ACS patients and a number of pets were transferred to the Cryonics Institute (CI) facility in Michigan.[4] The American Cryonics Society had contracted with CI for the long-term cryogenic storage of a number of other members prior to the 2002 patient transfer. The ACS inspects CI yearly to ensure ACS quality standards are met.[5] The extra funds charged to ACS members beyond CI minimums could be used for moving the patients in the future if necessary, or other uses.[5]

In 2004, the American Cryonics Society signed a contract with Suspended Animation, a Florida based company, and in the same year Suspended Animation performed a stand-by and suspension on an ACS member living in Florida at the time of his death. The patient was then transported to the CI Facility in Michigan for long-term cryogenic storage. When circumstances warrant, neural vitrification technology may also be applied to ACS subjects by CI personnel prior to long-term cryogenic storage.

3.2.2 Policies

The American Cryonics Society is managed by a seven person Board of Governors. Governors must themselves be full members, and are expected to have made suspension arrangements. Funds are managed by professional funds managers. As a further safeguard, a sponsor is designated for each person in suspension, to review financial records and cold-storage procedures.

The Cryonics Society of America works cooperatively with the Cryonics Institute and supports the research program of that organization. Most ACS (frozen) members are in suspension at the CI facility. The agreement with CI calls for the American Cryonics Society to have inspection authority, and to have the right to remove ACS subjects from that facility should the American Cryonics Society warrant that such removal is warranted and beneficial. Such subject would then be transferred to another facility. All ACS subjects are fully funded to CI's specification, and in addition, funds to benefit the subjects are maintained by the American Cryonics Society. This funding plan provides a double safety net against the possible financial insolvency of either CI or ACS.

The American Cryonics Society encourages members and the public in general to subscribe to the Immortalist Magazine. This bi-monthly magazine is devoted to discussion of cryonics activities, especially those of the Cryonics Institute where most patients of the American Cryonics Society are in long term cryogenic storage.

3.2.3 Scope

The American Cryonics Society welcomes members from any country, however the level of service that a member can expect in most countries outside the US and Canada is limited, and in many cases the member himself, sometimes working with other cryonics advocates, must work to make local arrangements. Even then, given so much uncertainty, the American Cryonics Society can make no guarantees.

3.2.4 Perspective

The cryonics movement, which started with the publication of Robert Ettinger's book *The Prospect of Immortality* in 1964, is still treated as a curiosity by most people, though cryonics continues to gain in acceptance by both the general public, and the scientific community.[6] Given the small number of people in suspension, and the incidents of patients lost to carelessness or accident in the past, the risk management approach to cryonics followed by the American Cryonics Society is warranted. That said, the extant cryonics organizations appear to have stability not always present in the past, and there have been no recent incidents of lost patients.

There are (perhaps) 1,800 people with pre-need cryonics arrangements in place; however many of these people are young and not apt to need suspension services for many years. As of the end of 2010, there are about 220 people now in suspension. These numbers include members and people in suspension from all organizations, in all countries. Almost half of them are at the Alcor facility, and half at the CI facility (which includes the patients of the American Cryonics Society). A couple of subjects are at Trans Time and KrioRus, and we will include in our count the dry ice storage of Mr. Bredo Morstoel, in Nederland, Colorado, though many cryonicists argue that because of the warmer storage temperature of Mr. Morstoel, he should not be counted as a "patient." Since the patients at the CI facility are almost all whole body patients (as opposed to head only), the volume of biological material at the CI facility is far greater than at the other facilities.

From 1974 to present, only one American Cryonics Society patient has had to be transferred out of long-term cold storage. In that case the transfer was following a request made by the patient's family, apparently for financial reasons. The patient was then preserved by chemical means. However, there have been several patients "converted" from whole-body to "neuro" because of funding concerns. The American Cryonics Society members either pre-approve or opt out of such arrangements as part of their contract. There have been no incidents in ACS history of embezzlement or loss of any patient funds.

3.2.5 See also

- Charles Platt (science-fiction author)
- Robert Ettinger

3.2.6 References

[1] Swank, Edgar (October 2010). "The top 13 reasons to join ACS". Retrieved 2011-02-11.

[2] Guynn, Jessica (2002). "Techies go for ice-cold afterlife". *Contra Costa Times*. Retrieved 2009-08-26.

[3] Mondragon, Carlos (1994). "Defining the Cryonics Institution". *Cryonics and Life Extension Conference*. Alcor Life Extension Foundation. Retrieved 2009-08-23.

[4] Best, Ben (2008). "A History of Cryonics" (PDF). *The Immortalist*. Cryonics Institute. Retrieved 2009-08-24.

[5] Best, Ben (2008). "Revival Assets Seminar" (PDF). *The Immortalist* (Cryonics Institute). Retrieved 2009-08-26.

[6] 61 Scientists. "Scientists' Open Letter on Cryonics". Retrieved 22 May 2012.

3.2.7 External links

- Official website

- Cryonics Institute

3.3 Immortalist Society

The **Immortalist Society** is a charitable 501(c)(3) organization devoted to research and education in the areas of cryonics and life extension. It was incorporated as a Michigan corporation by Robert Ettinger and five other local residents on June 27, 1967 as the *Cryonics Society of Michigan, Inc.*. In September 1976, the name of the corporation was changed to *Cryonics Association* in acknowledgement that its scope of operations was not limited to a single state. On October 20, 1985, the Articles of Incorporation were amended once more to change the name to *Immortalist Society*.

3.3.1 Organization

All Officers of the Immortalist Society are also Directors. As of May 2008, the Immortalist Society officers were:

3.3.2 Operation

Every two months the Immortalist Society publishes its flagship journal *Long Life: Longevity through Technology* (formerly *The Immortalist*), which is sent free to Members of the Cryonics Institute, but must be paid for by subscribers or Immortalist Society Members who do not join the Cryonics Institute. It is also available online for free. *Long Life* covers not only the activities of the Cryonics Institute, but activities of the American Cryonics Society and life extension news. Published six times per year, the magazine presents news, book reviews, technical articles, biographies, conference reports and other articles of interest to members.

The Immortalist Society is particularly supportive of the work of the Cryonics Institute. Donations to the Immortalist Society Research Fund are given to finance the research of Dr. Yuri Pichugin, the full-time Russian cryobiologist employed by the Cryonics Institute to develop vitrification mixture, improve perfusion protocol and find formulations to minimize cold ischemia (a concern for organ transplantation). Dr. Pichugin resigned from the Cryonics Institute in December 2007.[1] At the time of his resignation, the Cryonics Institute noted that Dr. Pichugin intended to work in Russia and continue his research for the Cryonics Institute and other interested organizations on a contract basis.

3.3.3 See also

- Immortality

3.3.4 References

[1] "Dr. Pichugin resigns, Chana de Wolf visits". Cryonics Institute. 2007-12-14. Retrieved 2008-05-01.

3.3.5 External links

- Cryonics Institute

- Immortalist Society

- Long Life magazine

3.4 KrioRus

KrioRus is the first Russian cryonics company. Established in 2005 by 8 Russian cryonicists, it is the first cryonics company outside the United States, except for Alcor-UK (formerly Mizar Limited).

KrioRus offers neurosuspension (cryopreservation of the brain) (US $12,000 as of January 2015), full-body cryonics suspension (US $36,000 as of January 2015), pet cryonics suspension, DNA storage, body's transportation in dry ice and other related services to clients from Russia, CIS and EU.[1][2]

The company stores the brains of several patients: Lidia Fedorenko, Lubov Chernaya, an anonymous 60-year-old man and others, in its facilities outside Moscow. Some full-body patients are also stored.

As of February 2013, KrioRus had 25 humans,[3] 3 cats, 5 dogs and 2 birds in cryopreservation. As of July 2015, KrioRus had 42 humans.[4][5]

3.4.1 See also

- Alcor Life Extension Foundation

- American Cryonics Society

- Cryonics Institute

3.4.2 References

[1] KrioRus Services

[2] Cryonics a reasoned choice of modern man

[3] KrioRus - Our patients

[4] KrioRus Services

[5] 2nd Cat Cryopreserved

3.4.3 External links

- KrioRus

- Publications about KrioRus

3.5 Life extension

Life extension science, also known as **anti-aging medicine**, **indefinite life extension**, **experimental gerontology**, and **biomedical gerontology**, is the study of slowing down or reversing the processes of aging to extend both the maximum and average lifespan. Some researchers in this area, and "life extensionists", "immortalists" or "longevists" (those who wish to achieve longer lives themselves), believe that future breakthroughs in tissue rejuvenation, stem cells, regenerative medicine, molecular repair, pharmaceuticals, and organ replacement (such as with artificial organs or xenotransplantations) will eventually enable humans to have indefinite lifespans (agerasia[1]) through complete rejuvenation to a healthy youthful condition.

The sale of putative anti-aging products such as nutrition, physical fitness, skin care, hormone replacements, vitamins, supplements and herbs is a lucrative global industry, with the US market generating about $50 billion of revenue each year.[2] Some medical experts state that the use of such products has not been proven to affect the aging process and many claims regarding the efficacy of these marketed products have been roundly criticized by medical experts, including the American Medical Association.[2][3][4][5][6]

However, it has not been shown that the goal of indefinite human lifespans itself is necessarily not feasible; some animals such as hydra, planarian flatworms, and certain sponges, corals, and jellyfish do not die of old age and exhibit potential immortality.[7][8][9][10] The ethical ramifications of life extension are debated by bioethicists.

3.5.1 Public opinion

Life extension is a controversial topic due to fear of overpopulation and possible effects on society.[11] Religious people are no more likely to oppose life extension than the unaffiliated,[12] though some variation exists between religious denominations. Biogerontologist Aubrey De Grey counters the overpopulation critique by pointing out that the therapy could postpone or eliminate menopause, allowing women to space out their pregnancies over more years and thus *decreasing* the yearly population growth rate.[13] Moreover, the philosopher and futurist Max More argues that, given the fact the worldwide population growth rate is slowing down and is projected to eventually stabilize and begin falling, superlongevity would be unlikely to contribute to overpopulation.[11]

A Spring 2013 Pew Research poll in the United States found that 38% of Americans would want life extension treatments, and 56% would reject it. However, it also found that 68% believed most people would want it and that only 4% consider an "ideal lifespan" to be more than 120 years. The median "ideal lifespan" was 91 years of age and the majority of the public (63%) viewed medical advances aimed at prolonging life as generally good. 41% of Americans believed that radical life extension (RLE) would be good for society, while 51% said they believed it would be bad for society.[12] One possibility for why 56% of Americans claim they would reject life extension treatments may be due to the cultural perception that living longer would result in a longer period of decrepitude, and that the elderly in our current society are unhealthy.[14]

3.5.2 Average and maximum lifespans

Main article: Senescence

During the process of aging, an organism accumulates damage to its macromolecules, cells, tissues, and organs. Specifically, aging is characterized as and thought to be caused by "genomic instability, telomere attrition, epigenetic alterations, loss of proteostasis, deregulated nutrient sensing, mitochondrial dysfunction, cellular senescence, stem cell exhaustion, and altered intercellular communication."[15] Oxidation damage to cellular contents caused by free radicals is believed to contribute to aging as well.[16][16][17]

The longest a human has ever been proven to live is 122 years, the case of Jeanne Calment who was born in 1875 and died in 1997, whereas the maximum lifespan of a wildtype mouse, commonly used as a model in research on aging, is about three years.[18] Genetic differences between humans and mice that may account for these different aging rates include differences in efficiency of DNA repair, antioxidant defenses, energy metabolism, proteostasis maintenance, and recycling mechanisms such as autophagy.[19]

Average lifespan in a population is lowered by infant and child mortality, which are frequently linked to infectious diseases or nutrition problems. Later in life, vulnerability to accidents and age-related chronic disease such as cancer or cardiovascular disease play an increasing role in mortality. Extension of expected lifespan can often be achieved by access to improved medical care, vaccinations, good diet, exercise and avoidance of hazards such as smoking.

Maximum lifespan is determined by the rate of aging for a species inherent in its genes and by environmental factors. Widely recognized methods of extending maximum lifespan in model organisms such as nematodes, fruit flies, and mice include caloric restriction, gene manipulation, and administration of pharmaceuticals.[20] Another technique uses evolutionary pressures such as breeding from only older members or altering levels of extrinsic mortality.[21][22]

Theoretically, extension of maximum lifespan in humans could be achieved by reducing the rate of aging damage by periodic replacement of damaged tissues, molecular repair or rejuvenation of deteriorated cells and tissues, reversal of harmful epigenetic changes, or the enhancement of telomerase enzyme activity.[23][24]

Research geared towards life extension strategies in various organisms is currently under way at a number of academic and private institutions. Since 2009, investigators have found ways to increase the lifespan of nematode worms and yeast by 10-fold; the record in nematodes was achieved through genetic engineering and the extension in yeast by a combination of genetic engineering and caloric restriction.[25] A 2009 review of longevity research noted: "Extrapolation from worms to mammals is risky at best, and it cannot be assumed that interventions will result in comparable life extension factors. Longevity gains from dietary restriction, or from mutations studied previously, yield smaller benefits to Drosophila than to nematodes, and smaller still to mammals. This is not unexpected, since mammals have evolved to live many times the worm's lifespan, and humans live nearly twice as long as the next longest-lived primate. From an evolutionary per-

spective, mammals and their ancestors have already undergone several hundred million years of natural selection favoring traits that could directly or indirectly favor increased longevity, and may thus have already settled on gene sequences that promote lifespan. Moreover, the very notion of a "life-extension factor" that could apply across taxa presumes a linear response rarely seen in biology."[25]

3.5.3 Current anti-aging strategies and issues

See also: Ageing § Prevention and reversal

Diets and supplements

Much life extension research focuses on nutrition—diets or supplements—as a means to extend lifespan, although few of these have been systematically tested for significant longevity effects. The many diets promoted by anti-aging advocates are often contradictory. A dietary pattern with some support from scientific research is caloric restriction.[26][27]

Preliminary studies of caloric restriction on humans using surrogate measurements have provided evidence that caloric restriction may have powerful protective effect against secondary aging in humans. Caloric restriction in humans may reduce the risk of developing Type 2 diabetes and atherosclerosis.[28] More research is needed.

The free-radical theory of aging suggests that antioxidant supplements, such as vitamin C, vitamin E, Q_{10}, lipoic acid, carnosine, and N-acetylcysteine, might extend human life. However, combined evidence from several clinical trials suggest that β-carotene supplements and high doses of vitamin E increase mortality rates.[29] Resveratrol is a sirtuin stimulant that has been shown to extend life in animal models, but the effect of resveratrol on lifespan in humans is unclear as of 2011.[30]

There are many traditional herbs purportedly used to extend the health-span, including a Chinese tea called Jiaogulan (*Gynostemma pentaphyllum*), dubbed "China's Immortality Herb."[31] Ayurveda, the traditional Indian system of medicine, describes a class of longevity herbs called rasayanas, including *Bacopa monnieri, Ocimum sanctum, Curcuma longa, Centella asiatica, Phyllanthus emblica, Withania somnifera* and many others.[31]

Hormone treatments

The anti-aging industry offers several hormone therapies. Some of these have been criticized for possible dangers to the patient and a lack of proven effect. For example, the American Medical Association has been critical of some anti-aging hormone therapies.[2]

Although some recent clinical studies have shown that low-dose growth hormone (GH) treatment for adults with GH deficiency changes the body composition by increasing muscle mass, decreasing fat mass, increasing bone density and muscle strength, improves cardiovascular parameters (i.e. decrease of LDL cholesterol), and affects the quality of life without significant side effects,[32][33][34] the evidence for use of growth hormone as an anti-aging therapy is mixed and based on animal studies. There are mixed reports that GH or IGF-1 signaling modulates the aging process in humans and about whether the direction of its effect is positive or negative.[35]

Scientific controversy regarding anti-aging nutritional supplementation and medicine

Some critics dispute the portrayal of aging as a disease. For example, Leonard Hayflick, who determined that fibroblasts are limited to around 50 cell divisions, reasons that aging is an unavoidable consequence of entropy. Hayflick and fellow biogerontologists Jay Olshansky and Bruce Carnes have strongly criticized the anti-aging industry in response to what they see as unscrupulous profiteering from the sale of unproven anti-aging supplements.[4]

Ethics and politics of anti-aging nutritional supplementation and medicine

Politics relevant to the substances of life extension pertain mostly to communications and availability.

In the United States, product claims on food and drug labels are strictly regulated. The First Amendment (freedom of speech) protects third-party publishers' rights to distribute fact, opinion and speculation on life extension practices. Manufacturers and suppliers also provide informational publications, but because they market the substances, they are subject to monitoring and enforcement by the Federal Trade Commission (FTC), which polices claims by marketers. What constitutes the difference between truthful and false claims is hotly debated and is a central controversy in this arena.

Consumer motivations for using anti-aging products

Research by Sobh and Martin (2011) suggests that people buy anti-aging products to obtain a hoped-for self (e.g., keeping a youthful skin) or to avoid a feared-self (e.g., looking old). The research shows that when consumers pur-

sue a hoped-for self, it is expectations of success that most strongly drive their motivation to use the product. The research also shows why doing badly when trying to avoid a feared self is more motivating than doing well. Interestingly, when product use is seen to fail it is more motivating than success when consumers seek to avoid a feared-self.[36]

3.5.4 Proposed strategies of life extension

Caloric restriction

The best-characterized anti-aging therapy was, and still is, CR. In some studies calorie restriction has been shown to extend the life of mice, yeast, and rhesus monkeys significantly.[37][38] However, a more recent study has shown that in contrast, calorie restriction has not improved the survival rate in rhesus monkeys.[39] Long-term human trials of CR are now being done. It is the hope of the anti-aging researchers that resveratrol, found in grapes, or pterostilbene, a more bio-available substance, found in blueberries, as well as rapamycin, a biotic substance discovered on Easter Island, may act as CR mimetics to increase the life span of humans.[40]

Anti-aging drugs

There are a number of chemicals intended to slow the aging process currently being studied in animal models. One type of research is related to the observed effects a calorie restriction (CR) diet, which has been shown to extend lifespan in some animals[41] Based on that research, there have been attempts to develop drugs that will have the same effect on the aging process as a caloric restriction diet, which are known as Caloric restriction mimetic drugs. Some drugs that are already approved for other uses have been studied for possible longevity effects on laboratory animals because of a possible CR-mimic effect; they include rapamycin,[42] and metformin.[43] Resveratrol and pterostilbene are dietary supplements that have also been studied in this context.[40][44][45]

Other attempts to create anti-aging drugs have taken different research paths. One notable direction of research has been research into the possibility of using the enzyme telomerase in order to counter the process of telomere shortening.[46] However, there are potential dangers in this, since some research has also linked telomerase to cancer and to tumor growth and formation.[47]

Nanotechnology

Future advances in nanomedicine could give rise to life extension through the repair of many processes thought to be responsible for aging. K. Eric Drexler, one of the founders of nanotechnology, postulated cell repair machines, including ones operating within cells and utilizing as yet hypothetical molecular computers, in his 1986 book Engines of Creation. Raymond Kurzweil, a futurist and transhumanist, stated in his book *The Singularity Is Near* that he believes that advanced medical nanorobotics could completely remedy the effects of aging by 2030.[48]

Cloning and body part replacement

Some life extensionists suggest that therapeutic cloning and stem cell research could one day provide a way to generate cells, body parts, or even entire bodies (generally referred to as reproductive cloning) that would be genetically identical to a prospective patient. Recently, the US Department of Defense initiated a program to research the possibility of growing human body parts on mice.[49] Complex biological structures, such as mammalian joints and limbs, have not yet been replicated. Dog and primate brain transplantation experiments were conducted in the mid-20th century but failed due to rejection and the inability to restore nerve connections. As of 2006, the implantation of bio-engineered bladders grown from patients' own cells has proven to be a viable treatment for bladder disease.[50] Proponents of body part replacement and cloning contend that the required biotechnologies are likely to appear earlier than other life-extension technologies.

The use of human stem cells, particularly embryonic stem cells, is controversial. Opponents' objections generally are based on interpretations of religious teachings or ethical considerations. Proponents of stem cell research point out that cells are routinely formed and destroyed in a variety of contexts. Use of stem cells taken from the umbilical cord or parts of the adult body may not provoke controversy.[51]

The controversies over cloning are similar, except general public opinion in most countries stands in opposition to reproductive cloning. Some proponents of therapeutic cloning predict the production of whole bodies, lacking consciousness, for eventual brain transplantation.

Cyborgs

Replacement of biological (susceptible on diseases) organs with mechanical ones could extend life. This is the goal of 2045 Initiative.[52]

Cryonics

Main article: Cryonics

For cryonicists (advocates of cryopreservation), storing the body at low temperatures after death may provide an "ambulance" into a future in which advanced medical technologies may allow resuscitation and repair. They speculate cryogenic temperatures will minimize changes in biological tissue for many years, giving the medical community ample time to cure all disease, rejuvenate the aged and repair any damage that is caused by the cryopreservation process.

Many cryonicists do not believe that legal death is "real death" because stoppage of heartbeat and breathing—the usual medical criteria for legal death—occur before biological death of cells and tissues of the body. Even at room temperature, cells may take hours to die and days to decompose. Although neurological damage occurs within 4–6 minutes of cardiac arrest, the irreversible neurodegenerative processes do not manifest for hours.[53] Cryonicists state that rapid cooling and cardio-pulmonary support applied immediately after certification of death can preserve cells and tissues for long-term preservation at cryogenic temperatures. People, particularly children, have survived up to an hour without heartbeat after submersion in ice water. In one case, full recovery was reported after 45 minutes underwater.[54] To facilitate rapid preservation of cells and tissue, cryonics "standby teams" are available to wait by the bedside of patients who are to be cryopreserved to apply cooling and cardio-pulmonary support as soon as possible after declaration of death.[55]

No mammal has been successfully cryopreserved and brought back to life, with the exception of frozen human embryos. Resuscitation of a postembryonic human from cryonics is not possible with current science. Some scientists still support the idea based on their expectations of the capabilities of future science.[56][57]

Strategies for Engineered Negligible Senescence (SENS)

Main articles: Strategies for Engineered Negligible Senescence and Genetics of aging

Another proposed life extension technology would combine existing and predicted future biochemical and genetic techniques. SENS proposes that rejuvenation may be obtained by removing aging damage via the use of stem cells and tissue engineering, removal of telomere-lengthening machinery, allotopic expression of mitochondrial proteins, targeted ablation of cells, immunotherapeutic clearance, and novel lysosomal hydrolases.[58]

While many biogerontologists find these ideas "worthy of discussion"[59][60] and SENS conferences feature important research in the field,[61][62] some contend that the alleged benefits are too speculative given the current state of technology, referring to it as "fantasy rather than science".[3][5]

Genetic modification

Gene therapy, in which nucleic acid polymers are delivered as a drug and are either expressed as proteins, interfere with the expression of proteins, or correct genetic mutations, has been proposed as a future strategy to prevent aging.[63][64]

A large array of genetic modifications have been found to increase lifespan in model organisms such as yeast, nematode worms, fruit flies, and mice. As of 2013, the longest extension of life caused by a single gene manipulation was roughly 150% in mice and 10-fold in nematode worms.[65]

Fooling genes

In *The Selfish Gene*, Richard Dawkins describes an approach to life-extension that involves "fooling genes" into thinking the body is young.[66] Dawkins attributes inspiration for this idea to Peter Medawar. The basic idea is that our bodies are composed of genes that activate throughout our lifetimes, some when we are young and others when we are older. Presumably, these genes are activated by environmental factors, and the changes caused by these genes activating can be lethal. It is a statistical certainty that we possess more lethal genes that activate in later life than in early life. Therefore, to extend life, we should be able to prevent these genes from switching on, and we should be able to do so by "identifying changes in the internal chemical environment of a body that take place during aging... and by simulating the superficial chemical properties of a young body".[67]

Reversal of informational entropy

According to some lines of thinking, the ageing process is routed into a basic reduction of biological complexity,[68] and thus loss of information. In order to reverse this loss, gerontologist Marios Kyriazis suggested that it is necessary to increase input of actionable and meaningful information both individually (into individual brains),[69] and collectively (into societal systems).[70] This technique enhances overall biological function through up-regulation of immune, hormonal, antioxidant and other parameters, resulting in improved age-repair mechanisms. Working in parallel with natural evolutionary mechanisms that can facilitate survival through increased fitness, Kryiazis claims that the technique may lead to a reduction of the rate of death as a function of age, i.e. indefinite lifespan.[71]

Mind uploading

Main article: Mind uploading

One hypothetical future strategy that, as some suggest, "eliminates" the complications related to a physical body, involves the copying or transferring (e.g. by progressively replacing neurons with transistors) of a conscious mind from a biological brain to a non-biological computer system or computational device. The basic idea is to scan the structure of a particular brain in detail, and then construct a software model of it that is so faithful to the original that, when run on appropriate hardware, it will behave in essentially the same way as the original brain.[72] Whether or not an exact copy of one's mind constitutes actual life extension is matter of debate.

3.5.5   History of the life extension movement

The extension of life has been a desire of humanity and a mainstay motif in the history of scientific pursuits and ideas throughout history, from the Sumerian Epic of Gilgamesh and the Egyptian Smith medical papyrus, all the way through the Taoists, Ayurveda practitioners, alchemists, hygienists such as Luigi Cornaro, Johann Cohausen and Christoph Wilhelm Hufeland, and philosophers such as Francis Bacon, René Descartes, Benjamin Franklin and Nicolas Condorcet. However, the beginning of the modern period in this endeavor can be traced to the end of the 19th – beginning of the 20th century, to the so-called "fin-de-siècle" (end of the century) period, denoted as an "end of an epoch" and characterized by the rise of scientific optimism and therapeutic activism, entailing the pursuit of life extension (or life-extensionism). Among the foremost researchers of life extension at this period were the Nobel Prize winning biologist Elie Metchnikoff (1845-1916) -- the author of the cell theory of immunity and vice director of Institut Pasteur in Paris, and Charles-Édouard Brown-Séquard (1817-1894) -- the president of the French Biological Society and one of the founders of modern endocrinology.[73]

Sociologist James Hughes claims that science has been tied to a cultural narrative of conquering death since the Age of Enlightenment. He cites Francis Bacon (1561–1626) as an advocate of using science and reason to extend human life, noting Bacon's novel *New Atlantis*, wherein scientists worked toward delaying aging and prolonging life. Robert Boyle (1627–1691), founding member of the Royal Society, also hoped that science would make substantial progress with life extension, according to Hughes, and proposed such experiments as "to replace the blood of the old with the blood of the young". Biologist Alexis Carrel (1873–1944) was inspired by a belief in indefinite human lifespan that he developed after experimenting with cells, says Hughes.[74]

In 1970, the American Aging Association was formed under the impetus of Denham Harman, originator of the free radical theory of aging. Harman wanted an organization of biogerontologists that was devoted to research and to the sharing of information among scientists interested in extending human lifespan.

In 1976, futurists Joel Kurtzman and Philip Gordon wrote *No More Dying. The Conquest Of Aging And The Extension Of Human Life*, (ISBN 0-440-36247-4) the first popular book on research to extend human lifespan. Subsequently, Kurtzman was invited to testify before the House Select Committee on Aging, chaired by Claude Pepper of Florida, to discuss the impact of life extension on the Social Security system.

Saul Kent published *The Life Extension Revolution* (ISBN 0-688-03580-9) in 1980 and created a nutraceutical firm called the Life Extension Foundation, a non-profit organization that promotes dietary supplements. The Life Extension Foundation publishes a periodical called *Life Extension Magazine*. The 1982 bestselling book *Life Extension: A Practical Scientific Approach* (ISBN 0-446-51229-X) by Durk Pearson and Sandy Shaw further popularized the phrase "life extension".

In 1983, Roy Walford, a life-extensionist and gerontologist, published a popular book called *Maximum Lifespan*. In 1988, Walford and his student Richard Weindruch summarized their research into the ability of calorie restriction to extend the lifespan of rodents in *The Retardation of Aging and Disease by Dietary Restriction* (ISBN 0-398-05496-7). It had been known since the work of Clive McCay in the 1930s that calorie restriction can extend the maximum lifespan of rodents. But it was the work of Walford and Weindruch that gave detailed scientific grounding to that knowledge. Walford's personal interest in life extension motivated his scientific work and he practiced calorie restriction himself. Walford died at the age of 80 from complications caused by amyotrophic lateral sclerosis.

Money generated by the non-profit Life Extension Foundation allowed Saul Kent to finance the Alcor Life Extension Foundation, the world's largest cryonics organization. The cryonics movement had been launched in 1962 by Robert Ettinger's book, *The Prospect of Immortality*. In the 1960s, Saul Kent had been a co-founder of the Cryonics Society of New York. Alcor gained national prominence when baseball star Ted Williams was cryonically preserved by Alcor in 2002 and a family dispute arose as to whether Williams had really wanted to be cryopreserved.

Regulatory and legal struggles between the Food and Drug Administration (FDA) and the Life Extension Foundation included seizure of merchandise and court action. In 1991, Saul Kent and Bill Faloon, the principals of the Foundation, were jailed. The LEF accused the FDA of perpetrating a "Holocaust" and "seeking gestapo-like power" through its regulation of drugs and marketing claims.[75]

In 2003, Doubleday published "The Immortal Cell: One Scientist's Quest to Solve the Mystery of Human Aging," by Michael D. West. West emphasised the potential role of embryonic stem cells in life extension.[76]

Other modern life extensionists include writer Gennady Stolyarov, who insists that death is "the enemy of us all, to be fought with medicine, science, and technology";[77] transhumanist philosopher Zoltan Istvan, who proposes that the "transhumanist must safeguard one's own existence above all else";[78] futurist George Dvorsky, who considers aging to be a problem that desperately needs to be solved;[79] and recording artist Steve Aoki, who has been called "one of the most prolific campaigners for life extension".[80]

Scientific research

In 1991, the American Academy of Anti-Aging Medicine (A4M) was formed as a non-profit organization to create what it considered an anti-aging medical specialty distinct from geriatrics, and to hold trade shows for physicians interested in anti-aging medicine. The A4M trains doctors in anti-aging medicine and publicly promotes the field of anti-aging research. It has about 26,000 members, of whom about 97% are doctors and scientists.[81] The American Board of Medical Specialties recognizes neither anti-aging medicine nor the A4M's professional standing.[82]

In 2003, Aubrey de Grey and David Gobel formed the Methuselah Foundation, which gives financial grants to anti-aging research projects. In 2009, de Grey and several others founded the SENS Research Foundation, a California-based scientific research organization which conducts research into aging and funds other anti-aging research projects at various universities.[83] In 2013, Google announced Calico, a new company based in San Francisco that will harness new technologies to increase scientific understanding of the biology of aging.[84] It is led by Arthur D. Levinson,[85] and its research team includes scientists such as Hal V. Barron, David Botstein, and Cynthia Kenyon. In 2014, biologist Craig Venter founded Human Longevity Inc., a company dedicated to scientific research to end aging through genomics and cell therapy. They received funding with the goal of compiling a comprehensive human genotype, microbiome, and phenotype database.[86]

Aside from private initiatives, aging research is being conducted in university laboratories, and includes universities such as Harvard and UCLA. University researchers have made a number of breakthroughs in extending the lives of mice and insects by reversing certain aspects of aging.[87][88][89][90]

3.5.6 Ethics and politics of life extension

Though many scientists state[91] that life extension and radical life extension are possible, there are still no international or national programs focused on radical life extension. There are political forces staying for and against life extension. By 2012, in Russia, the United States, Israel, and the Netherlands, the Longevity political parties started. They aimed to provide political support to radical life extension research and technologies, and ensure the fastest possible and at the same time soft transition of society to the next step – life without aging and with radical life extension, and to provide access to such technologies to most currently living people.[92]

Leon Kass (chairman of the US President's Council on Bioethics from 2001 to 2005) has questioned whether potential exacerbation of overpopulation problems would make life extension unethical.[93] He states his opposition to life extension with the words:

> "simply to covet a prolonged life span for ourselves is both a sign and a cause of our failure to open ourselves to procreation and to any higher purpose ... [The] desire to prolong youthfulness is not only a childish desire to eat one's life and keep it; it is also an expression of a childish and narcissistic wish incompatible with devotion to posterity."[94]

John Harris, former editor-in-chief of the Journal of Medical Ethics, argues that as long as life is worth living, according to the person himself, we have a powerful moral imperative to save the life and thus to develop and offer life extension therapies to those who want them.[95]

Transhumanist philosopher Nick Bostrom has argued that any technological advances in life extension must be equitably distributed and not restricted to a privileged few.[96] In an extended metaphor entitled "The Fable of the Dragon-Tyrant", Bostrom envisions death as a monstrous dragon who demands human sacrifices. In the fable, after a lengthy debate between those who believe the dragon is a fact of life and those who believe the dragon can and should be destroyed, the dragon is finally killed. Bostrom argues that political inaction allowed many preventable human deaths to occur.[97]

3.5.7 Aging as a disease

Most mainstream medical organizations and practitioners do not consider aging to be a disease. David Sinclair says: "I don't see aging as a disease, but as a collection of quite predictable diseases caused by the deterioration of the body".[98] The two main arguments used are that aging is both inevitable and universal while diseases are not.[99] However, not everyone agrees. Harry R. Moody, Director of Academic Affairs for AARP, notes that what is normal and what is disease strongly depends on a historical context.[100] David Gems, Assistant Director of the Institute of Healthy Ageing, strongly argues that aging should be viewed as a disease.[101] In response to the universality of aging, David Gems notes that it is as misleading as arguing that Basenji are not dogs because they do not bark.[102] Because of the universality of aging he calls it a 'special sort of disease'. Robert M. Perlman, coined the terms 'aging syndrome' and 'disease complex' in 1954 to describe aging.[103]

The discussion whether aging should be viewed as a disease or not has important implications. It would stimulate pharmaceutical companies to develop life extension therapies and in the United States of America, it would also increase the regulation of the anti-aging market by the FDA. Anti-aging now falls under the regulations for cosmetic medicine which are less tight than those for drugs.[102][104]

3.5.8 See also

Main articles: List of life extension topics and Index of life extension-related articles

- Advanced glycation end product

- Aging

- Aging brain

- Aging movement control

- Alzheimer's disease

- Anti-aging movement

- Centenarian

- *Clinical Interventions in Aging*

- Dementia

- DNA damage theory of aging

- Human enhancement

- Immortality

- Maximum lifespan
- *Rejuvenation Research*
- Senescence
- Slow aging
- Supercentenarian
- Transgenerational design

3.5.9 References

[1] "agerasia". *Oxford English Dictionary* (3rd ed.). Oxford University Press. September 2005. (Subscription or UK public library membership required.)

[2] Japsen, Bruce (15 June 2009). "AMA report questions science behind using hormones as anti-aging treatment". *The Chicago Tribune*. Retrieved 17 July 2009.

[3] Holliday, Robin (2008). "The extreme arrogance of anti-aging medicine". *Biogerontology* **10** (2): 223–8. doi:10.1007/s10522-008-9170-6. PMID 18726707.

[4] Olshansky, S. J.; Hayflick, L; Carnes, B. A. (1 August 2002). "Position statement on human aging". *The Journals of Gerontology Series A: Biological Sciences and Medical Sciences* **57** (8): B292–7. doi:10.1093/gerona/57.8.B292. PMID 12145354.

[5] "Science fact and the SENS agenda. What can we reasonably expect from ageing research?". *EMBO Reports* **6** (11): 1006–8. 2005. doi:10.1038/sj.embor.7400555. PMC 1371037. PMID 16264422.

[6] Marziali, Carl (7 December 2010). "Reaching Toward the Fountain of Youth". *USC Trojan Family Magazine*. Retrieved 7 December 2010.

[7] Newmark, P. A.; Sánchez Alvarado, A (2002). "Not your father's planarian: a classic model enters the era of functional genomics". *Nat Rev Genet* **3** (3): 210–219. doi:10.1038/nrg759. PMID 11972158.

[8] Bavestrello, G.; Sommer, C.; Sarà, M. (1992). "Bidirectional conversion in *Turritopsis nutricula* (Hydrozoa)" (PDF). *Scientia Marina* **56** (2–3): 137–140.

[9] Martínez DE (May 1998). "Mortality patterns suggest lack of senescence in hydra". *Experimental Gerontology* **33** (3): 217–25. doi:10.1016/S0531-5565(97)00113-7. PMID 9615920.

[10] Petralia, Ronald S.; Mattson, Mark P.; Yao, Pamela J. (2014). "Aging and longevity in the simplest animals and the quest for immortality". *Ageing Res Rev* **16**: 66–82. doi:10.1016/j.arr.2014.05.003. PMC 4133289. PMID 24910306.

[11] "Superlongevity Without Overpopulation". *Fight Aging!*.

[12] "Living to 120 and Beyond: Americans' Views on Aging, Medical Advances and Radical Life Extension". *Pew Research Center's Religion & Public Life Project*. 6 August 2013.

[13] "Peter Singer on Should We Live to 1,000? – Project Syndicate". *Project Syndicate*.

[14] de Magalhães JP (2014). "The scientific quest for lasting youth: prospects for curing aging". *Rejuvenation Res* **17** (5): 458–67. doi:10.1089/rej.2014.1580. PMC 4203147. PMID 25132068.

[15] López-Otín, C; Blasco, M. A.; Partridge, L; Serrano, M; Kroemer, G (2013). "The hallmarks of aging". *Cell* **153** (6): 1194–1217. doi:10.1016/j.cell.2013.05.039. PMC 3836174. PMID 23746838.

[16] Halliwell B, Gutteridge JMC (2007). Free Radicals in Biology and Medicine. Oxford University Press, USA, ISBN 019856869X, ISBN 978-0198568698

[17] Holmes, G. E.; Bernstein, C; Bernstein, H (September 1992). "Oxidative and other DNA damages as the basis of aging: a review". *Mutation Research/DNAging* **275** (3–6): 305–15. doi:10.1016/0921-8734(92)90034-M. PMID 1383772.

[18] "Mouse Facts". informatics.jax.org.

[19] "What Causes Aging? Damage-Based Theories of Aging".

[20] Verdaguer, E; Junyent, F; Folch, J; Beas-Zarate, C; Auladell, C; Pallàs, M; Camins, A (2012). "Aging biology: a new frontier for drug discovery". *Expert Opin Drug Discov* **7** (3): 217–229. doi:10.1517/17460441.2012.660144. PMID 22468953.

[21] Rauser, C. L.; Mueller, L. D.; Rose, M. R. (2006). "The evolution of late life". *Ageing Res Rev.* **5** (1): 14–32. doi:10.1016/j.arr.2005.06.003. PMID 16085467.

[22] Stearns, S. C.; Ackermann, M; Doebeli, M; Kaiser, M (2000). "Experimental evolution of aging, growth, and reproduction in fruitflies". *Proceedings of the National Academy of Sciences of the United States of America* **97** (7): 3309–3313. doi:10.1073/pnas.060289597. PMC 16235. PMID 10716732.

[23] Rando TA; Chang HY (2012). "Aging, rejuvenation, and epigenetic reprogramming: resetting the aging clock". *Cell* **148** (1–2): 46–57. doi:10.1016/j.cell.2012.01.003. PMID 22265401.

[24] Johnson AA, Akman K, Calimport SR, Wuttke D, Stolzing A, de Magalhães JP; Akman; Calimport; Wuttke; Stolzing; De Magalhães (2012). "The role of DNA methylation in aging, rejuvenation, and age-related disease". *Rejuvenation Res* **15** (5): 483–494. doi:10.1089/rej.2012.1324. PMC 3482848. PMID 23098078.

[25] Shmookler Reis, R. J.; Bharill, P; Tazearslan, C; Ayyadevara, S (2009). "Extreme-longevity mutations orchestrate silencing of multiple signaling pathways". *Biochim Biophys Acta* **1790** (10): 1075–83. doi:10.1016/j.bbagen.2009.05.011. PMC 2885961. PMID 19465083.

[26] Schumacher, B; van der Pluijm, I; Moorhouse, MJ (2008). "Delayed and Accelerated Aging Share Common Longevity Assurance Mechanisms". *PLoS Genetics* **4** (8): e1000161. doi:10.1371/journal.pgen.1000161. PMC 2493043. PMID 18704162.

[27] Chen, J; Velalar, CN; Ruan, R (2008). "Identifying the changes in gene profiles regulating the amelioration of age-related oxidative damages in kidney tissue of rats by the intervention of adult-onset calorie restriction". *Rejuvenation Research* **11** (4): 757–63. doi:10.1089/rej.2008.0718. PMID 18710334.

[28] Holloszy, J. O.; Fontana, L. (2007). "Caloric restriction in humans. Experimental gerontology,". *Experimental Gerontology* **42** (8): 707–712. doi:10.1016/j.exger.2007.03.009. PMID 17482403.

[29] Bjelakovic, Goran; Nikolova, Dimitrinka; Lotte Gluud, Lise; Simonetti Rosa G.; Gluud Christian (2007). "Mortality in Randomized Trials of Antioxidant Supplements for Primary and Secondary Prevention, a Systematic Review and Meta-analysis". *JAMA* **297** (8): 842–857. doi:10.1001/jama.297.8.842. PMID 17327526.

[30] Fernández AF; Fraga MF (Jul 2011). "The effects of the dietary polyphenol resveratrol on human healthy aging and lifespan". *Epigenetics : official journal of the DNA Methylation Society* **6** (7): 870–4. doi:10.4161/epi.6.7.16499. PMID 21613817.

[31] Mishra, R. N.; Joshi, D. (2011). "Jiao Gu Lan (*Gynostemma pentaphyllum*): The Chinese Rasayan" (PDF). *International Journal of Research in Pharmaceutical and Biomedical Sciences*.

[32] Alexopoulou, O; Abs, R; Maiter, D (2010). "Treatment of adult growth hormone deficiency: who, why and how? A review". *Acta Clinica Belgica* **65** (1): 13–22. doi:10.1179/acb.2010.002. PMID 20373593.

[33] Ahmad, A. M.; Hopkins, M. T.; Thomas, J; Ibrahim, H; Fraser, W. D.; Vora, J. P. (June 2001). "Body composition and quality of life in adults with growth hormone deficiency; effects of low-dose growth hormone replacement". *Clinical Endocrinology* **54** (6): 709–17. doi:10.1046/j.1365-2265.2001.01275.x. PMID 11422104.

[34] Savine R; Sönksen P (2000). "Growth hormone – hormone replacement for the somatopause?". *Hormone Research* **53** (Suppl 3): 37–41. doi:10.1159/000023531. PMID 10971102.

[35] Sattler FR (August 2013). "Growth hormone in the aging male". *Best Pract. Res. Clin. Endocrinol. Metab.* **27** (4): 541–55. doi:10.1016/j.beem.2013.05.003. PMID 24054930. In animal models, alterations in GH/IGF-1 signaling with reductions in these somatotrophs appear to increase life span. ... Administration of IGF-1Eb (mechanogrowth factor) stimulates proliferation of myoblasts and induces muscle hypertrophy. Increases in GH and IGF-1 during adolescence are beneficial for brain and cardiovascular function during the aging process and GH administration during adolescence is vasoprotective and increases life-span.[15] ... Studies relating GH and IGF-1 status to longevity provide inconsistent evidence as to whether decreased (somatopause) or high levels (e.g. acromegaly) of these hormones are beneficial or detrimental to longevity. ... It is difficult to reconcile the largely protective effects of GH/IGF-1 deficiency on longevity in animals with the inconsistent or deleterious effects of low levels or declining GH/IGF-1 during human aging.

[36] "Feedback Information and Consumer Motivation. The Moderating Role of Positive and Negative Reference Values in Self-Regulation" (PDF). *European Journal of Marketing* **45** (6): 963–986. 2011. doi:10.1108/03090561111119976.

[37] "Metabolic and behavioral compensations in response to caloric restriction: implications for the maintenance of weight loss". *PLOS ONE* **4** (2): e4377. 2009. doi:10.1371/journal.pone.0004377.

[38] Holloszy JO; Fontana L (2007). "Caloric restriction in humans". *Exp Gerontol* **42** (8): 709–12. doi:10.1016/j.exger.2007.03.009. PMID 17482403.

[39] "Impact of caloric restriction on health and survival in rhesus monkeys from the NIA study". *Nature* **489** (7415): 318–321. 2012. doi:10.1038/nature11432. PMC 3832985. PMID 22932268.

[40] Kaeberlein, Matt (2010). "Resveratrol,pterostilbene and rapamycin:are they anti-aging drugs?". *BioEssays* **32** (2): 96–99. doi:10.1002/bies.200900171. PMID 20091754.

[41] Anderson, M.; Shanmuganayagam, D.; Weindruch, R. (2009). "Caloric restriction and aging: studies in mice and monkeys". *Toxicologic pathology* **37** (1): 47–51. doi:10.1177/0192623308329476. PMID 19075044.

[42] "Rapamycin fed late in life extends lifespan in genetically heterogeneous mice". *Nature* **460**: 392–5. 2009. doi:10.1038/nature08221. PMC 2786175. PMID 19587680.

[43] "Identification of Potential Caloric Restriction Mimetics by Microarray Profiling". *Physiological Genomics* **23** (3): 343–50. 2005. doi:10.1152/physiolgenomics.00069.2005. PMID 16189280.

[44] "A Low Dose of Dietary Resveratrol Partially Mimics Caloric Restriction and Retards Aging Parameters in Mice". *PLOS ONE* **3** (6): e2264. doi:10.1371/journal.pone.0002264. PMC 2386967. PMID 18523577.

[45] McCormack D, McFadden D. A review of pterostilbene antioxidant activity and disease modification. Oxid Med Cell Longev. 2013;2013:575482. PMID 23691264 PMC 3649683

[46] "Telomeres and Telomerase Basic Science Implications for Aging". *American Geriatrics Society* **49** (8): 1105–1109. doi:10.1046/j.1532-5415.2001.49217.x.

[47] Blackburn, E. H. (2005). "Telomerase and Cancer: Kirk A. Landon - AACR Prize for Basic Cancer Research Lecture". *Molecular Cancer Research* **3** (9): 477–82. doi:10.1158/1541-7786.MCR-05-0147. PMID 16179494.

[48] Kurzweil, Ray (2005). *The Singularity Is Near*. New York City: Viking Press. ISBN 978-0-670-03384-3. OCLC 57201348.

[49] Melanson, Donald (April 22, 2008). "DoD establishes institute tasked with regrowing body parts". Engadget. Retrieved June 29, 2010.

[50] Khamsi, Roxanne (April 4, 2006). "Bio-engineered bladders successful in patients". New Scientist. Retrieved January 26, 2011.

[51] White, Christine (19 August 2005). "Umbilical stem cell breakthrough". *The Australian*. Retrieved 17 July 2009.

[52] David Segal for the New York Times. 1 June 2013 This Man Is Not a Cyborg. Yet.

[53] "Neuronal necrosis after middle cerebral artery occlusion in Wistar rats progresses at different time intervals in the caudoputamen and the cortex". *Stroke* **26** (4): 636–42; discussion 643. 1995. doi:10.1161/01.STR.26.4.636. PMID 7709411.

[54] "Full recovery after 45 min accidental submersion". *Intensive Care Medicine* **28** (4): 524. April 2002. doi:10.1007/s00134-002-1245-2. PMID 11967613.

[55] "Comprehensive Member Standby". Retrieved 14 December 2010.

[56] "Scientists' Open Letter on Cryonics". Retrieved 17 July 2009.

[57] "Advances in Cryonics". Retrieved 14 December 2010.

[58] de Grey, Aubrey; Michael Rae (2007). *Ending Aging: The Rejuvenation Breakthroughs that Could Reverse Human Aging in Our Lifetime*. New York City: St. Martin's Press. ISBN 978-0-312-36706-0. OCLC 132583222.

[59] Pontin, Jason (July 11, 2006). "Is Defeating Aging Only A Dream?". *Technology Review*.

[60] Garreau, Joel (October 31, 2007). "Invincible Man". *Washington Post*.

[61] Fourth SENS Conference (2009). Over 140 Accepted Abstracts. Cambridge, England, September 3–7th, 2009.

[62] Kristen Fortney (2009). SENS4 Conference Coverage From Ouroboros. FightAging.org, September 4, 2009.

[63] Goya, Rodolfo G.; Federico Bolognani; Claudia B. Hereñú; Omar J. Rimoldi (2001-01-08). "Neuroendocrinology of Aging: The Potential of Gene Therapy as an Interventive Strategy". *Gerontology* **47** (168–173): 168–173. doi:10.1159/000052792.

[64] Rattan, S. I. S.; Singh, R. (2008-10-22). "Progress & Prospects: Gene therapy in aging". *Gene Therapy* **16** (3–9): 3–9. doi:10.1038/gt.2008.166. PMID 19005494.

[65] Tacutu, R.; Craig, T.; Budovsky, A.; Wuttke, D.; Lehmann, G.; Taranukha, D.; Costa, J.; Fraifeld, V. E.; De Magalhaes, J. P. (2012). "Human Ageing Genomic Resources: Integrated databases and tools for the biology and genetics of ageing". *Nucleic Acids Research* **41** (Database issue): D1027–33. doi:10.1093/nar/gks1155. PMC 3531213. PMID 23193293.

[66] Dawkins, Richard (2006) [1976]. *The Selfish Gene*. New York: Oxford University Press. pp. 41–42. ISBN 978-0-19-929115-1.

[67] Dawkins, Richard (2006) [1976]. *The Selfish Gene*. New York: Oxford University Press. p. 42. ISBN 978-0-19-929115-1.

[68] Lipsitz, L. A. (2006). "Aging as a Process of Complexity Loss". *Complex Systems Science in Biomedicine*. Topics in Biomedical Engineering International Book Series. p. 641. doi:10.1007/978-0-387-33532-2_28. ISBN 978-0-387-30241-6.

[69] "h+ Magazine – The Longevity of Real Human Avatars – h+ Magazine". *h+ Magazine*.

[70] Kyriazis, Marios (October 30, 2012). "The Myth of the Longevity Elixir". ieet.org.

[71] Kryiazis, Marios (December 2, 2012) The Global Brain and its Role in Human Immortality. immortallife.info

[72] Sandberg, Anders; Boström, Nick (2008). *Whole Brain Emulation: A Roadmap* (PDF). Technical Report #2008-3. Future of Humanity Institute, Oxford University. Retrieved 7 March 2013. *The basic idea is to take a particular brain, scan its structure in detail, and construct a software model of it that is so faithful to the original that, when run on appropriate hardware, it will behave in essentially the same way as the original brain.*

[73] Stambler, Ilia (2014). *A History of Life-Extensionism in the Twentieth Century*. Longevity History. ISBN 1500818577.

[74] Hughes, James (October 20, 2011). "Transhumanism". In Bainbridge, William. *Leadership in Science and Technology: A Reference Handbook*. Sage Publications. p. 587. ISBN 1452266522.

[75] Clevenger, Ty (Summer 2000). "Internet pharmacies: cyberspace versus the regulatory state". *Journal of Law and Health*. Retrieved 17 July 2009.

[76] West, Michael D. (2003). *The Immortal Cell: One Scientist's Quest to Solve the Mystery of Human Aging*. Doubleday. ISBN 978-0-385-50928-2.

[77] Stolyarov, Gennady (November 25, 2013). *Death is Wrong* (PDF). Rational Argumentator Press. ISBN 978-0615932040.

[78] Istvan, Zoltan (October 2, 2014). "The Morality of Artificial Intelligence and the Three Laws of Transhumanism". Huffington Post.

[79] "Futurist: 'I will reap benefits of life extension'". *Al Jazeera America*. May 7, 2015. To Dvorsky, aging is a problem that's desperately in need of solving.

[80] Tez, Riva Melissa (May 11, 2015). "Steve Aoki, Dan Bilzerian, a giraffe and the search for eternal life". *i-D*. VICE. Unknown to most, Steve is both an undeniable champion of life expansion as well as one of the most prolific campaigners for life extension. Understanding that the depth of his life's experience is limited by time alone, in his latest album Neon Future he pens lyrics such as 'Life has limitless variety... But today, because of ageing, it does not have limitless scope.' [...] Set up by the Steve Aoki Charitable Fund, the profits from the Dan Bilzerian party went to life extension research.

[81] "About the A4M". *Worldhealth.net*.

[82] Kuczynski, Alex (12 April 1998). "Anti-Aging Potion Or Poison?". *The New York Times*. Retrieved 17 July 2009.

[83] research report 2011. Sens Foundation

[84] Arion McNicoll, Arion (3 October 2013). "How Google's Calico aims to fight aging and 'solve death'". *CNN*.

[85] "Google announces Calico, a new company focused on health and well-being". Google. September 18, 2013.

[86] Human Longevity Inc. (4 March 2014). "Human Longevity Inc. (HLI) Launched to Promote Healthy Aging Using Advances in... – SAN DIEGO, March 4, 2014 /PRNewswire/ --".

[87] Landau, Elizabeth (5 May 2014). "Young blood makes old mice more youthful". *CNN*.

[88] "Harvard researchers find protein that could reverse the aging process". *gizmag.com*.

[89] Wolpert, Stuart. "UCLA biologists delay the aging process by 'remote control'". UCLA.edu.

[90] "Australian and US scientists reverse ageing in mice, humans could be next". *ABC News*.

[91] "Scientists' Open Letter on Aging". Imminst.org. Retrieved 2012-10-07.

[92] "A Single-Issue Political Party for Longevity Science". Fightaging.org. Retrieved 2012-10-07.

[93] Smith, Simon (3 December 2002). "Killing Immortality". Betterhumans. Archived from the original on 7 June 2004. Retrieved 17 July 2009.

[94] Kass, Leon (1985). *Toward a more natural science: biology and human affairs*. New York City: Free Press. p. 316. ISBN 978-0-02-918340-3. OCLC 11677465.

[95] Harris J. (2007) *Enhancing Evolution: The ethical case for making better people*. Princeton University Press, New Jersey.

[96] Sutherland, John (9 May 2006). "The ideas interview: Nick Bostrom". *The Guardian* (London). Retrieved 17 July 2009.

[97] Bostrom, N (May 2005). "The fable of the dragon tyrant". *Journal of Medical Ethics* **31** (5): 273–7. doi:10.1136/jme.2004.009035. PMC 1734155. PMID 15863685.

[98] Hayden EC (2007). "A new angle on 'old'". *Nature* **450** (7170): 603–603. doi:10.1038/450603a. PMID 18046373.

[99] Hamerman D. (2007) *Geriatric Bioscience: The link between aging & disease*. The Johns Hopkins University Press, Maryland.

[100] Moody HR (2002). "Who's afraid of life extension?". *Generations* **25** (4): 33–7.

[101] Gems D (2011). "Aging: To Treat, or Not to Treat? The possibility of treating aging is not just an idle fantasy". *American Scientist* **99** (4): 278–80. doi:10.1511/2011.91.278.

[102] Gems D (2011). "Tragedy and delight: the ethics of decelerated ageing". *Phil Trans R Soc B.* **366** (1561): 108–112. doi:10.1098/rstb.2010.0288.

[103] Perlman RM (1954). "The aging syndrome". *J Am Geriatr Soc.* **2**: 123–129.

[104] Mehlman, M. J.; Binstock, R. H.; Juengst, E. T.; Ponsaran, R. S.; Whitehouse, P. J. (2004). "Anti-aging medicine: Can consumers be better protected?". *The Gerontologist* **44** (3): 304–10. PMID 15197284.

3.5.10 External links

- Greg Easterbrook, "What Happens When We All Live to 100?, *The Atlantic*, October 2014, pp. 60–72.

- Ezekiel J. Emanuel, Why I Hope to Die at 75, *The Atlantic*, October 2014, pp. 74–81.

- Aubrey de Grey: 'We will be able to live to 1,000'

3.6 Life Extension Society

Not to be confused with Life Extension Foundation.

The **Life Extension Society** (**LES**) with its network of co-ordinators was the first cryonics organization in the world. It was founded by Evan Cooper in 1964 to promote cryonic suspension of people, and became the seed tree for cryonics societies throughout the USA where local cryonics advocates would meet as a result of contact through the LES mailing list. The original LES ceased existence near the end of the 1960s, but an organization with the same name and similar objectives was incorporated in Maryland in 1992.[1]

3.6.1 History

In 1962, Cooper privately published a manuscript named *Immortality: Physically, Scientifically, Now* under his pseudonym, Nathan Duhring. The book is considered by Michael C. Price "a modest, almost apologetic one; the ideas it contains are the stuff of genius and the fabric of change, in it he advocated that men need not be born only to die and that if they were frozen at or near the time of death they might yet have a chance to live again, whole and complete, forever." [2]

In the same year, but shortly after Cooper's book appeared, a Michigan college physics teacher, Robert Ettinger, privately published his book *The Prospect of Immortality*, that independently suggested the same idea. Ettinger came to be credited as the originator of cryonics, perhaps because his book was republished by Doubleday in 1964 on the recommendations of Isaac Asimov and Fred Pohl, and received more publicity. Ettinger also stayed with the movement longer. Nevertheless, the cryonics historian R. Michael Perry has written: "Evan Cooper deserves the principal credit for forming an organized cryonics movement."[3]

Cooper stopped his cryonics activities by 1970. His former wife, Milred, said that "he turned away from cryonics because of overload, burn-out, and a general sense that it was not going to be a viable option in his lifetime" and possibly because there was little or no actual scientific support of its methods. The remaining time of his life he spent sailing, until he was lost at sea in 1983.

3.6.2 See also

- Life extension

3.6.3 References

[1] "Life Extension Society". *A Not-for-Profit Maryland Corpo-*

ration, Incorporated 1992. keithlynch.net. Retrieved 2014-02-13.

[2] Ev Cooper,*Cryonics*, March 1983, accessed 13 June 2013

[3] Michael Perry,"Unity and Disunity in Cryonics", *Cryonics*, Volume 13(8) Issue 145, pg 5, August 1992, accessed 13 June 2013

3.6.4 External links

- Evan Cooper, *Immortality: Physically, Scientifically, Now*

- Robert Ettinger, *The Prospect of Immortality*

3.7 Suspended Animation, Inc

Suspended Animation, Inc (**SA**) was founded in 2002 in Boynton Beach, FL.[1][2] SA's purpose is to preserve bodies immediately after legal death to minimize the damages that occur before the body is cryopreserved. SA does not actually perform final cryopreservation, rather, they work with companies such as Alcor Life Extension Foundation and Cryonics Institute which carry out the cryopreservations. Unlike Alcor Life Extension Foundation and Cryonics Institute, Suspended Animation, Inc does not offer memberships,[3] but rather gains revenue from performing the one-time procedure.[4]

3.7.1 Coordination with Cryonics Organizations

Cryonics Institute (CI) offers coordination with Suspended Animation, Inc for patients outside of Michigan (CI's headquarters). CI also expresses that it "neither endorses nor opposes the use of SA" and urges members to make informed decisions regarding the use of Suspended Animation, Inc.[5]

Alcor Life Extension Foundation also offers options for patients to use Suspended Animation, Inc in coordination with their procedures.[6] Suspended Animation, Inc. provides all standby/stabilization/transport services for terminal Alcor Members outside Arizona, but inside the continental United States. Alcor provides those services for terminal Alcor Members in Canada, Arizona, and Hawaii.[7]

In 2011, Suspended Animation hosted a weekend conference on Cryonics.[8]

3.7.2 References

[1] Sun Sentinel, 2004 Cryonics Company Targets Boynton

[2] , Suspended Animation.

[3] , fightaging.org.

[4] , Suspended Animation: FAQ.

[5] , Cryonics Institute.

[6] , Alcor.

[7] Best, Ben.   "Cryonics Services Offered".   *Cryonics*.
    Longecity. Retrieved 2013-08-13.

[8] Alex Beam, May 2011 A cold day in Florida

Chapter 4

Related Articles to Cryonics

4.1　21st Century Medicine

For events in medicine in the 21st century, see 21st century#Medicine.

21st Century Medicine (21CM) is a California cryobiological research company which has as its primary focus the development of perfusates and protocols for viable long-term cryopreservation of human organs, tissues and cells at cryogenic temperatures (temperatures below −100 °C) through the use of vitrification. 21CM was founded in 1993.

Dr. Gregory M. Fahy, who pioneered the use of vitrification in reproductive cryopreservation,[1][2][3][4][5][6][7][8][9] serves on the company's Board of Directors and prioritizes, develops and directs the company's research activities. He also manages all extramural collaborative research projects with universities, industry and research institutions to create specific products and services.

The company holds a number of patents, most notably for cryoprotectant mixtures that greatly reduce ice formation while minimizing cryoprotectant toxicity, as well as for synthetic ice-blockers that inexpensively simulate the antifreeze protein found in arctic organisms. Their website lists peer-reviewed journal publications based on research conducted in their laboratories.[10] In 2004 21CM received a $900,000 grant from the U.S. National Institutes of Health (NIH) to study a preservation solution developed by the University of Rochester in New York for extending simple cold storage time of human hearts removed for transplant.[11]

At the July 2005 annual conference of the Society for Cryobiology, 21st Century Medicine announced the vitrification of a rabbit kidney to −135 °C with their proprietary vitrification mixture. The kidney was successfully transplanted upon rewarming to a rabbit, the rabbit being euthanized on the 48th day for histological follow-up.[12][13]

4.1.1　See also

- Cryobiology
- Cryopreservation
- Biobank
- Tissue Engineering
- Organ Transplantation

4.1.2　References

[1] Michael J. Taylor, Ying C. Song, and Kelvin G.M. Brockbank (2004). *Vitrification in Tissue Preservation: New Developments In:* Life in the Frozen State *(B.J. Fuller, N. Lane, and E.E. Benson, Eds.).* CRC Press. pp. 603–641. ISBN 0-415-24700-4.

[2] Fahy, G.M. and Hirsh, A. (1982). *Prospects for Organ Preservation by Vitrification. In:* Organ Preservation, Basic and Applied Aspects *(D.E. Pegg, I.A. Jacobsen and N.A. Halasz, Eds.).* Springer. pp. 399–404. ISBN 0-85200-418-4.

[3] Fahy GM, MacFarlane DR, Angell CA, Meryman HT (1984). "Vitrification as an approach to cryopreservation". *Cryobiology* **21** (4): 407–426. doi:10.1016/0011-2240(84)90079-8. PMID 6467964.

[4] Rall WF, Fahy GM (1985). "Ice-free cryopreservation of mouse embryos at −196 degrees C by vitrification". *Nature (journal)* **313** (6003): 573–575. doi:10.1038/313573a0. PMID 3969158.

[5] Fahy GM (1986). "Vitrification: a new approach to organ cryopreservation". *PROGRESS IN CLINICAL AND BIOLOGICAL RESEARCH* **224**: 305–335. PMID 3540994.

[6] Fahy, Gregory M. (May 16, 2002). "Vitrification versus Freezing of Organs". *Science (journal) E-Letter responses.* American Association for the Advancement of Science. Retrieved 2010-10-22.

51

[7] Fahy, G.M., and Rall, W.F (2007). *Vitrification: An overview. In:* Vitrification in Assisted Reproduction: A User's Manual and Troubleshooting Guide *(J. Liebermann and M.J. Tucker, Eds).* Informa Healthcare. ISBN 0-415-40882-2.

[8] Mullen, S.F., and Fahy, G.M (February 28, 2011). *Fundamental aspects of vitrification as a method of reproductive cell, tissue, and organ cryopreservation. In:* Principles & Practice of Fertility Preservation *(Donnez, J., and Kim, S.S., Eds.).* Cambridge University Press. ISBN 0-521-19695-7.

[9] "Cryopreservation of embryos". *The Lancet* **1** (8430): 678. 1985. doi:10.1016/s0140-6736(85)91336-4. PMID 2858625.

[10] "21CM Publications". Archived from the original on 18 October 2006. Retrieved 2006-11-08.

[11] "NIH grant to be used for heart preservation research". Business Wire. October 31, 2004. Retrieved 2012-05-02.

[12] "Plenary Session: Fundamentals of Biopreservation". *CRYO 2005 Scientific Program.* Society for Cryobiology. July 24, 2005. Archived from the original on 2006-08-30. Retrieved 2006-11-08.

[13] Fahy GM, Wowk B, Pagotan R, Chang A, Phan J, Thomson B, Phan L (2009). "Physical and biological aspects of renal vitrification". *ORGANOGENESIS* **5** (3): 167–175. doi:10.4161/org.5.3.9974. PMC 2781097. PMID 20046680. After ensuring that the animal appeared capable of living indefinitely using the vitrified kidney as the sole renal support, it was euthanized for histological follow-up on day 48.

4.1.3　External links

- 21st Century Medicine

4.2　Chemical brain preservation

Chemical brain preservation is the proposed process of using aldehyde fixation for long term storage of a brain with the intent of future revival. It would be considered an alternative or adjunct to cryonics.

4.2.1　Technology

The Brain Preservation Foundation is offering the Brain Preservation Technology Prize in order to promote research and development in the field. As of March 2015, the prize is valued at $106,840.[1]

Vascular perfusion of a brain with chemical fixative agents followed by a plasticizing agent is one possible approach.

This plastic embedding is widely used to study small sections ($<1 \text{mm}^2$) of human and animal brain tissue under laboratory conditions.

4.2.2　The future

It is unclear at present how much of the mind can be rescued from a preserved brain, irrespective of the preservation techniques used. Pioneers of this new technology might have themselves preserved based on the reasonable assumption that at least some of the information that defines their mind is preserved, and that technology will continue to progress, and at some point in the future, procedures will be developed that allow the information which defines the mind to be separated from the "noise" caused by present day preservation efforts.[2]

A potential but unverified (in humans) alternative to physical revival of the brain is digital emulation of the biological computation process of the brain. The preserved physical brain could potentially be scanned by special microscopes, and then the relevant structure is re-created as a computer model, which could then "run" to recreate the biological process of thinking. (This is also known as a brain simulation). Such may potentially make it easier to "repair" damage to the physical specimen due to age or the preservation process because it's a digital copy(s). However, just like physical revival, emulation of a complete human mind is still in a highly speculative state typically served by science fiction writers and philosophers.

4.2.3　See also

- Connectome

- Chemical fixation

- Plastination

- Suspended animation

- Transhumanism

4.2.4　References

[1] "Technology Prize". *TECHNOLOGY PRIZE.* The Brain Preservation Foundation. June 12, 2010. Retrieved 2015-06-27.

[2] http://cryonics.org/probability.html

- Bachofen, H., Ammann, A., Wangensteen, D., & Weibel, E.R. (1982). Perfusion fixation of lungs for structure-function analysis: credits and limitations. J Appl Physiol, 53 (2), 528-533.

- Denk, W. and H. Horstmann (2004). Serial block-face scanning electron microscopy to reconstruct three-dimensional tissue nanostructure. PLoS Biol 2(11): e329

- Fahy, G. M., B. Wowk, et al. (2004). Cryopreservation of organs by vitrification: perspectives and recent advances. Cryobiology 48(2): 157-78.

- Hayworth, K. J., N. Kasthuri, R. Schalek and J. W. Lichtman. 2006. Automating the Collection of Ultrathin Serial Sections for Large Volume TEM Reconstructions. Extended abstract of a paper presented at Microscopy and Microanalysis 2006 in Chicago, Illinois, USA, July 30 – August 3, 2006

- Hayworth, K. J. (2007) ATLUM Project Home Page. http://www.mcb.harvard.edu/lichtman/ATLUM/ATLUM_web.htm

- Knott, G., Marchman, H., Wall, D., & Lich, B. (2008). Serial Section Scanning Electron Microscopy of Adult Brain Tissue Using Focused Ion Beam Milling. The Journal of Neuroscience, 28 (12), 2959–2964.

- Krucker, T., Lang, A., & Meyer, E.P. (2006). New polyurethane-based material for vascular corrosion casting with improved physical and imaging characteristics. Microsc Res Tech, 69 (2), 138-147.

- Kurzweil, R. (2006). The Singularity is Near. Penguin press

- Lemler, J., Harris, S. B., Platt, C., & Huffman, T. M. (2004). The arrest of biological time as a bridge to engineered negligible senescence. Ann N Y Acad Sci, 1019, 559-563.

- Markram, H. (2006). The blue brain project. Nat Rev Neurosci 7(2): 153-60.

- Oldmixon, E.H., Suzuki, S., Butler, J.P., & Hoppin, F.G., Jr. (1985). Perfusion dehydration fixes elastin and preserves lung air-space dimensions. J Appl Physiol, 58 (1), 105-113.

- Palay, S. L., McGee-Russell, S. M., Gordon, S., Grillo, M. A. (1962). Fixation of neural tissues for electron microscopy by perfusion with solutions of osmium tetroxide. J Cell Biol, 12, 385-410.

- Pichugin, Y., G. M. Fahy, et al. (2006). Cryopreservation of rat hippocampal slices by vitrification. Cryobiology 52(2): 228-40.

- Sandberg, A., and Bostrom, N. (2008). Whole Brain Emulation: A Roadmap. Technical Report #2008-3, Future of Humanity Institute, Oxford University.

- Sullivan, B. J., L. N. Sekhar, et al. (1999). "Profound hypothermia and circulatory arrest with skull base approaches for treatment of complex posterior circulation aneurysms." Acta Neurochir (Wien) 141(1): 1-11

4.2.5 External links

- Biostasis through chemopreservation
- The Brain Preservation Prize
- The Permafrost Papers

4.3 Extropianism

Extropianism, also referred to as the philosophy of *Extropy*, is an evolving framework of values and standards for continuously improving the human condition. Extropians believe that advances in science and technology will some day let people live indefinitely. An extropian may wish to contribute to this goal, e.g. by doing research and development or volunteering to test new technology.

Extropianism describes a pragmatic consilience of transhumanist thought guided by a proactionary approach to human evolution and progress.

Originated by a set of principles developed by Dr. Max More, *The Principles of Extropy*,[1] extropian thinking places strong emphasis on rational thinking and practical optimism. According to More, these principles "do not specify particular beliefs, technologies, or policies". Extropians share an optimistic view of the future, expecting considerable advances in computational power, life extension, nanotechnology and the like. Many extropians foresee the eventual realization of indefinite lifespans, and the recovery, thanks to future advances in biomedical technology or mind uploading, of those whose bodies/brains have been preserved by means of cryonics.

4.3.1 Extropy

The term 'extropy', as an antonym to 'entropy' was used in a 1967 academic volume discussing cryogenics[2] and in a 1978 academic volume of cybernetics.[3] Diane Duane was the first to use the term "extropy" to signify a potential transhuman destiny for humanity.[4] 'Extropy' as coined by Tom Bell (T.O. Morrow) and defined by Max More in 1988, is "the extent of a living or organizational system's intelligence, functional order, vitality, energy, life, experience, and capacity and drive for improvement and growth." Extropy is not a rigorously defined technical term in philosophy or science; in a metaphorical sense, it simply expresses the opposite of entropy.

A more recent definition of Extropy has been provided by Kevin Kelly, senior maverick at Wired magazine.[5] "Extropy is neither wave, nor particle, nor pure energy. It is a non-material force that is very much like information. Since Extropy is defined as negative entropy-the reversal of disorder-it is, by definition, an increase in order." Kelly gives this definition of extropy in his research on the evolution of technology.

In the philosophy of digital probabilistic physics, the extropy of a physical system is defined to be the self-information of the Markov chain probability of the physical system at a moment in time. This was to distinguish the probability of the Markov state of the physical system from the probability defined by entropy which creates ensembles of equivalent microstates.

4.3.2   The Extropy Institute

In 1987, Max More moved to Los Angeles from Oxford University in England, where he had helped to establish (along with Michael Price, Garret Smyth and Luigi Warren) the first European cryonics organization, known as Mizar Limited (later Alcor UK), to work on his Ph.D. in philosophy at the University of Southern California.

In 1988, *Extropy: The Journal of Transhumanist Thought* was first published. (For the first few issues, it was "Extropy: Vaccine for Future Shock.") This brought together thinkers with interests in artificial intelligence, nanotechnology, genetic engineering, life extension, mind uploading, idea futures, robotics, space exploration, memetics, and the politics and economics of transhumanism. Alternative media organizations soon began reviewing the magazine, and it attracted interest from like-minded thinkers. Later, More and Bell co-founded the Extropy Institute, a non-profit 501(c)(3) educational organization. "ExI" was formed as a transhumanist networking and information center to use current scientific understanding along with critical and creative thinking to define a small set of principles or values that could help make sense of new capabilities opening up to humanity.

The Extropy Institute's email list was launched in 1991 (and, as of April 2015, continues to exist as "Extropy-Chat"), and in 1992 the institute began producing the first conferences on transhumanism. Affiliate members throughout the world began organizing their own transhumanist groups. Extro Conferences, meetings, parties, online debates, and documentaries continue to spread transhumanism to the public.

The Internet soon became the most fertile breeding ground

for people interested in exploring transhumanist ideas, with the availability of websites for such organizations that have joined the Extropy Institute in developing and advocating transhumanist (and related) ideas. These include Humanity+, the Alcor Life Extension Foundation, the Life Extension Foundation, Foresight Institute, Transhumanist Arts & Culture, Betterhumans, the Singularity Institute for Artificial Intelligence, and the Institute for Ethics and Emerging Technologies.

In 2006, the board of directors of the Extropy Institute made a decision to close the organisation, stating that its mission was "essentially completed."[6]

4.3.3   Extropism

Extropism is a modern derivative of the transhumanist philosophy of Extropianism. It follows in the same tradition, hence the similarity of name, but has been revised to better suit the paradigms of the 21st century. As introduced in *The Extropist Manifesto*,[7] it promotes an optimistic futuristic philosophy that can be summed up in the following five phrases, which spell out the word "EXTROPISM":

- Endless eXtension
- Transcending Restriction
- Overcoming Property
- Intelligence
- Smart Machines

These five key points, when taken together, formulate a philosophy and world view which embraces bio-ethical abolitionism, life extension, singularitarianism, technogaianism, freedom of information and several other related disciplines and philosophies. While it does not make a firm political stance, it is most closely related to libertarian socialism (given that it supports the abolition of money and property). Philosophically, it draws from the philosophy of Jeremy Bentham and utilitarianism.

Extropists desire to prolong their life span to a near-immortal state and exist in a world where artificial intelligence and robotics have made work irrelevant. As in utilitarianism, the purpose of one's life should be to increase the overall happiness of all creatures on Earth through cooperation.

The Extropist Manifesto, written by Breki Tomasson and Hank Hyena of The Extropist Examiner in January 2010 (site since discontinued), details the ways in which Extropism has evolved away from, while building upon the original tenets of Extropianism. For example, it moves away from the original Extropian Principles[8] by placing a significant focus on the need to abolish and/or restrict the current use of surveillance, copyright and patent laws. This philosophy, inspired in part by the philosophy of the International Pirate Party, is one of the five basic tenets of the Extropist philosophy, falling under the category "Overcoming Property". Other noteworthy topics that appear frequently in Extropist writings is the focus on equal rights for LGBT couples and individuals and a general distaste for organized religiosity.

4.3.4 See also

- Biopunk movement

- Cyborg anthropology

- Democratic transhumanism

- Digital probabilistic physics

- Eclipse Phase (role-playing game), a tabletop game which uses the philosophy in its futuristic setting.

- Futures studies

- Holism

- Law of Complexity/Consciousness

- Negentropy

- Proactionary Principle

- Sustainability

- Systems philosophy

- Systems thinking

- Transhumanism

4.3.5 References

[1] Max More (2003). "Principles of Extropy (Version 3.11) : An evolving framework of values and standards for continuously improving the human condition". Extropy Institute. Archived from the original on 2013-10-15

[2] *Cryogenics*, IPC Science and Technology Press, vol. 7, pg. 225 (1967)

[3] Proceedings of the Fourth International Congress of Cybernetics & Systems: "Current Topics in Cybernetics and Systems", pg. 258 (1978)

[4] Duane, Diane. "The Wounded Sky" (1983)

[5] Kelly, Kevin (April 2011). "Understanding Technological Evolution and Diversity". *The Futurist* **45** (2): 44–48. ISSN 0016-3317.

[6] Extropy Institute (2006). "Next Steps". Retrieved 2006-05-05.

[7] The Extropist Manifesto. *The Extropist Examiner* (blog).

[8] Max More (1998). "The Extropian Principles (Version 3.0) : A Transhumanist Declaration". Extropy Institute.

4.3.6 External links

- Kevin Kelly on Extropy - Kevin Kelly at The Technium, August 29, 2009

- "Transhumanism's Extropy Institute - Transhumanism for a better future". Retrieved 1 August 2013.

4.4 Hibernation

For the power management process in computing, see hibernation (computing).
"Hibernate" redirects here. For the Java database library, see Hibernate (Java).

Hibernation is a state of inactivity and metabolic de-

Northern bat hibernating in Norway

pression in endotherms. Hibernation refers to a season of heterothermy that is characterized by low body temperature, slow breathing and heart rate, and low metabolic rate. Although traditionally reserved for "deep" hibernators such as rodents, the term has been redefined to include animals such as bears and is now applied based on active metabolic suppression [1] rather than based on absolute body temperature decline. Many experts believe that the processes of daily torpor and hibernation form a continuum and utilize similar mechanisms.[2] Hibernation during the summer months is known as aestivation. Some reptile species (ectotherms) are said to brumate, or undergo brumation, but any possible similarities between brumation and hibernation are not firmly established. Some insects, such as the wasp *Polistes exclamans* hibernate by aggregating together in groups in protected places called hibernacula.[3]

Often associated with low temperatures, the function of hibernation is to conserve energy during a period when sufficient food is unavailable. To achieve this energy saving, an endotherm will first decrease its metabolic rate, which then results in a decreased body temperature.[2] Hibernation may last several days, weeks, or months depending on the species, ambient temperature, time of year, and individual's body condition.

Before entering hibernation, animals need to store enough energy to last through the entire winter. Larger species become hyperphagic and eat a large amount of food and store the energy in fat deposits. In many small species, food caching replaces eating and becoming fat.[4] Some species of mammals hibernate while gestating young, which are either born while the mother hibernates or shortly afterwards.[5]

For example, the female polar bear goes into hibernation during the cold winter months to give birth to her offspring. She loses 15-27% of her pre-hibernation weight and uses stored fats for energy during times of food scarcity, or hibernation. It is evident that pregnant female polar bears significantly increase body mass prior to hibernation, and this increase is further reflected in the weight of their offspring. The fat accumulation prior to hibernation in female polar bears enables them to provide a sufficient and warm, nurturing environment for their newborns.[6]

Alternately, the term hibernation may commonly refer to the science-fiction concept of prolonged or indefinite suspended animation of humans or other organisms.

4.4.1 Hibernating animals

Primates

While hibernation has long been studied in rodents, namely ground squirrels, no primate or tropical mammal was known to hibernate prior to the discovery that the fat-tailed dwarf lemur of Madagascar hibernates in tree holes for seven months of the year.[7] Malagasy winter temperatures sometimes rise to over 30 °C (86 °F), so hibernation is not exclusively an adaptation to low ambient temperatures. The hibernation of this lemur is strongly dependent on the thermal behaviour of its tree hole: if the hole is poorly insulated, the lemur's body temperature fluctuates widely, passively following the ambient temperature; if well insulated, the body temperature stays fairly constant and the animal undergoes regular spells of arousal.[8] Dausmann found that hypometabolism in hibernating animals is not necessarily coupled to a low body temperature.[9]

Bears

For many decades it remained controversial whether bears actually hibernated, because over-wintering bears only experienced a modest drop in core-body temperature compared to smaller animals. What defines hibernation, however, is not the degree of temperature reduction, but the metabolic suppression. Adult bears can, however, lower metabolic rate to some 75% below basal metabolic rates, which indicates that bears are indeed hibearnators. Indeed, northern-most bears will neither eat nor drink for periods as long as 8 months, relying only on stored body-fat reserves for energy and water. Though it is believed that bear hibernation is very different from either rodent or primate hibernation and involves temperature-independent metabolic suppression, because the modest decreases in core temperature do not account for the large decrease in metabolic rate, this belief does not consider the effect of metabolic reductions that can occur through extensive peripheral vasoconstriction. For example, it is known that peripheral tissues contribute as much as 50% to metabolism. This discrepancy alone would be sufficient to account for the `missing´ proportion and without having to resort to more esoteric physiologic mechanism. This effect has been observed in other torpid metabolic states, like diving. In diving penguins and seals, for example, metabolic rate can be lowered without resorting to any core (visceral) temperature decreases merely through extensive vasoconstriction of peripheral tissue beds.

They are able to recycle their proteins and urine, allowing them to both stop urinating for months and stop muscle atrophy.[10][11]

Note that in some languages a specific term is used to describe the type of hibernation undergone by bears. In French for example it is called "hivernation" (not to be confused with "hivernage" which means "overwintering") instead of "hibernation".

Obligate hibernators

Obligate hibernators are defined as animals that sponta-neously, and annually, enter hibernation regardless of am-bient temperature and access to food. Obligate hiber-nators include many species of ground squirrels, other rodents, mouse lemurs, the European hedgehog and other insectivores, monotremes, marsupials, and even butterflies such as the small tortoiseshell.[12] These undergo what has been traditionally called "hibernation": the physiological state where the body temperature drops to near ambi-ent (environmental) temperature, and heart and respiration rates slow drastically. The typical winter season for these hibernators is characterized by periods of torpor inter-rupted by periodic, euthermic arousals, wherein body tem-peratures and heart rates are restored to euthermic (more typical) levels. The cause and purpose of these arousals is still not clear.

The question of why hibernators may experience the pe-riodic arousals (returns to high body temperature) has plagued researchers for decades, and while there is still no clear-cut explanation, there are myriad hypotheses on the topic. One favored hypothesis is that hibernators build a 'sleep debt' during hibernation, and so must occasionally warm up in order to sleep. This has been supported by evidence in the Arctic ground squirrel.[13] Another theory states that the brief periods of high body temperature dur-ing hibernation are used by the animal to restore its avail-able energy sources.[14] Yet another theory states that the frequent returns to high body temperature allow mammals to initiate an immune response.[15]

Hibernating Arctic ground squirrels may exhibit abdominal temperatures as low as −2.9 °C, maintaining sub-zero ab-dominal temperatures for more than three weeks at a time, although the temperatures at the head and neck remain at 0 °C or above.[16]

Historically there was a question of whether or not bears truly hibernate, since they experience only a modest decline in body temperature (3–5 K) compared with what other hibernators undergo (32 K or more). Many researchers thought that their deep sleep was not comparable with true, deep hibernation. This theory has been refuted by recent research in captive black bears.[17]

Facultative hibernation

Unlike obligate hibernators, facultative hibernators only en-ter hibernation when either cold stressed or food deprived, or both. A good example of the differences between the two types of hibernation can be seen among the prairie dogs: the white-tailed prairie dog is an obligate hibernator and the closely related black-tailed prairie dog is a facultative

Black bear mother and cubs "denning"

hibernator.[18]

4.4.2 Hibernating birds

Historically, Pliny the Elder believed swallows hibernated, and ornithologist Gilbert White pointed to anecdotal ev-idence in *The Natural History of Selborne* that indicated as much. Birds typically do not hibernate, instead utiliz-ing torpor. One known exception is the common poor-will (*Phalaenoptilus nuttallii*), first documented by Edmund Jaeger.[19][20]

4.4.3 Dormancy in fish

Fish are ectothermic, and so, by definition, cannot hiber-nate because they cannot actively down-regulate their body temperature or their metabolic rate. However, they can ex-perience decreased metabolic rates associated with colder environments and/or low oxygen availability (hypoxia) and can experience dormancy. For a couple of generations dur-ing the 20th century it was thought that basking sharks settled to the floor of the North Sea and became dor-mant. Research by Dr David Sims in 2003 dispelled this hypothesis,[21] showing that the sharks actively traveled huge distances throughout the seasons, tracking the areas with the highest quantity of plankton. The epaulette sharks have been documented to be able to survive for long periods of time without oxygen, even being left high and dry, and at temperatures of up to 26 °C (79 °F).[22] Other animals able to survive long periods with no or very little oxygen include the goldfish, the red-eared slider turtle, the wood frog, and the bar-headed goose.[23] However, the ability to survive hypoxic or anoxic conditions is not the same, nor closely related, to endotherm hibernation.

4.4.4 Hibernation induction trigger

Hibernation induction trigger (HIT) is a bit of misnomer. Although research in the 1990s hinted at the ability to induce torpor in animals by injection of blood taken from a hibernating animal, further research has been unable to reproduce this phenomenon. Despite the inability to induce torpor, there are substances in hibernator blood that can lend protection to organs for possible transplant. Researchers were able to prolong the life of an isolated pig's heart with a HIT.[24] This may have potentially important implications for organ transplant, as it could allow organs to survive for up to 18 or more hours, outside the human body. This would be a great improvement from the current 6 hours.

This supposed HIT is a mixture derived from serum, including at least one opioid-like substance. DADLE is an opioid that in some experiments has been shown to have similar functional properties.[25]

4.4.5 Human hibernation

See also: Suspended animation

There are many research projects currently investigating how to achieve "induced hibernation" in humans.[26][27] This ability to hibernate humans would be useful for a number of reasons, such as saving the lives of seriously ill or injured people by temporarily putting them in a state of hibernation until treatment can be given.

Actual and anecdotal cases of suspected human hibernation or states similar to hibernation exist in the literature:

- **Anna Bågenholm**, a Swedish radiologist who in 1999 survived 80 minutes under ice in a frozen stream in Norway, the final 40 minutes in a state of cardiac arrest, and survived with no brain damage.

- **Mitsutaka Uchikoshi**, a Japanese man who survived the cold for 24 days in 2006 without food or water when he fell into a state similar to hibernation[28]

- **Paulie Hynek**, who, at age 2, survived several hours of hypothermia-induced cardiac arrest and whose body temperature reached 64 °F (18 °C)[29]

- **John Smith**, a 14-year-old boy who survived 15 minutes under ice in a frozen lake before paramedics arrived to pull him onto dry land and saved him.[30]

4.4.6 See also

- Cryobiology

- Karolina Olsson, the "Sleeping Beauty of Oknö"

4.4.7 References

[1] Watts PD, Oritsland NA, Jonkel C, Ronald K (1981). "Mammalian hibernation and the oxygen consumption of a denning black bear (Ursus americanus)". *Comparative Biochemistry and Physiology Part A: Physiology* **69** (1): 121–3. doi:10.1016/0300-9629(81)90645-9.

[2] Geiser, Fritz (2004). "Metabolic Rate and Body Temperature Reduction During Hibernation and Daily Torpor". *Annu. Rev. Physiol.* **66**: 239–274. doi:10.1146/annurev.physiol.66.032102.115105.

[3] J.M. Gonzalez, and S.B. Vinson, "Does Polistes exclamans Vierek (Hymenoptera: Vespidae) Hibernate Inside Muddauber Nests," Southwestern Entomologist, vol. 32, no. 1, pp. 67-71, 2007. < http://www.bioone.org/doi/full/10.3958/0147-1724-32.1.69>

[4] Humphries, M. M.; Thomas, D.W.; Kramer, D.L. (2003). "The role of energy availability in mammalian hibernation: A cost-benefit approach". *Physiological and Biochemical Zoology* **76** (2): 165–179. doi:10.1086/367950.

[5] Hellgren, Eric C. (1998). "Physiology of Hibernation in Bears". *Ursus* **10**: 467–477. JSTOR 3873159.

[6] Molnar, PK, Derocher, AE, Kianjscek, T, Lewis, MA. Predicting climate change impacts on polar bear litter size. Nat Comm, 2:186, 2011.

[7] Dausmann, K.H.; Glos, J.; Ganzhorn, J.U.; Heldmaier, G. (2005). "Hibernation in the tropics: lessons from a primate". *Comparative Physiology B* **175** (3): 147–155. doi:10.1007/s00360-004-0470-0.

[8] Blanco, M. B.; Dausmann, K.; Ranaivoarisoa, J. F.; Yoder, A. D. (2013). "Underground Hibernation in a Primate". *Scientific Reports*. doi:10.1038/srep01768.

[9] "Physiology: Hibernation in a tropical primate" **429** (6994).

[10] Lundberg, D.A.; Nelson, R.A.; Wahner, H.W.; Jones, J.D. (1976). "Protein metabolism in the black bear before and during hibernation". *Mayo Clinnic Proceedings* **51** (11): 716–722.

[11] Nelson, R.A. (1980). "Protein and fat metabolism in hibernating bears". *FASEB J.* **39** (12): 2955–2958. PMID 6998737.

[12] Territorial Behaviour of the Nymphalid Butterflies, Aglais urticae (L.) and Inachis io (L.) R. R. Baker Journal of Animal Ecology , Vol. 41, No. 2 (Jun., 1972), pp. 453-469

[13] Daan S, Barnes BM, Strijkstra AM (1991). "Warming up for sleep? Ground squirrels sleep during arousals from hibernation". *Neurosci. Lett.* **128** (2): 265–8. doi:10.1016/0304-3940(91)90276-Y. PMID 1945046.

[14] Galster, W.; Morrison, P.R. (1975). "Gluconeogenesis in arctic ground squirrels between periods of hibernation". *American Journal of Physiology* **228** (1): 325–330.

[15] Prendergast, B.J.; Freeman, D.A.; Zucker, I.; Nelson, R.J. (2002). "Periodic arousal from hibernation is necessary for initiation of immune responses in ground squirrels". *AJP - Regu. Physiol.* **282** (4): R1054–R1062. doi:10.1152/ajpregu.00562.2001. PMID 11893609.

[16] Barnes, Brian M. (30 June 1989). "Freeze Avoidance in a Mammal: Body Temperatures Below 0 °C in an Arctic Hibernator" (PDF). *Science* (American Association for the Advancement of Science) **244** (4912): –1616. doi:10.1126/science.2740905. PMID 2740905. Retrieved 2008-11-23.

[17] Toien, Oivind; Black, J.; Edgar, D.M.; Grahn, D.A.; Heller, H.C.; Barnes, B.M. (February 2011). "Black Bears: Independence of Metabolic Suppression from temperature". *Science* **331** (6019): 906–909. doi:10.1126/science.1199435. PMID 21330544.

[18] Harlow, H.J.; Frank, C.L. (2001). "The role of dietary fatty acids in the evolution of spontaneous and facultative hibernation patterns in prairie dogs". *J. Comp. Physiol. B.* **171**: 77–84. doi:10.1007/s003600000148.

[19] Jaeger, Edmund C. (May–June 1949). "Further Observations on the Hibernation of the Poor-will". *The Condor.* 3 **51**: 105–109. JSTOR 1365104. Earlier I gave an account (Condor, 50, 1948:45) of the behavior of a Poor-will (*Phalaenoptilus nuttallinii*) which I found in a state of profound torpidity in the winter of 1946-47 in the Chuckawalla Mountains of the Colorado Desert, California.

[20] McKechnie, Andrew W.; Ashdown, Robert A. M., Christian, Murray B. & Brigham, R. Mark. "Torpor in an African caprimulgid, the freckled nightjar *Caprimulgus tristigma*" (PDF). *Journal of Avian Biology* **38** (3): 261–266. doi:10.1111/j.2007.0908-8857.04116.x.

[21] "Seasonal movements and behavior of basking sharks from archival tagging". *Marine Ecology Progress Series* **248**: 187–196. 2003. doi:10.3354/meps248187.

[22] "A Shark With an Amazing Party Trick". *New Scientist* **177** (2385): 46. 8 March 2003. Retrieved 2006-10-06.

[23] Breathless: A shark with an amazing party trick is teaching doctors how to protect the brains of stroke patients. Douglas Fox, New Scientist vol 177 issue 2385 — 8 March 2003, page 46. Last accessed November 9, 2006.

[24] Bolling, S.F.; Tramontini, N.L., Kilgore, K.S., Su, T-P., Oeltgen, P.R., Harlow, H.H. (1997). "Use of "Natural" Hibernation Induction Triggers for Myocardial Protection". *The Annals of Thoracic Surgery* **64** (3): 623–627. doi:10.1016/s0003-4975(97)00631-0.

[25] Oeltgen PR, Nilekani SP, Nuchols PA, Spurrier WA, Su TP (1988). "Further studies on opioids and hibernation: delta opioid receptor ligand selectively induced hibernation in summer-active ground squirrels". *Life Sc.* **43** (19): 1565–74. doi:10.1016/0024-3205(88)90406-7. PMID 2904105.

[26] New Hibernation Technique might work on humans | LiveScience at www.livescience.com

[27] Race to be first to 'hibernate' human beings - Times Online at www.timesonline.co.uk

[28] Japanese man in mystery survival at BBC News

[29] *Eleva boy's story part of national tour to honor Mayo Clinics 150 years* Mayo Clinic

[30] Suspended Animation? How A Boy Survived 15 Minutes Trapped Under Ice In Frozen Lake at Medical Daily

4.4.8 Further reading

- Carey, H.V., M.T. Andrews and S.L. Martin. 2003. Mammalian hibernation: cellular and molecular responses to depressed metabolism and low temperature. Physiological Reviews 83: 1153-1181.

4.4.9 External links

- Hibernaut

- Do Black Bears Hibernate?

- Freeze avoidance in a Mammal: Body Temperatures Below 0°C in an Arctic Hibernator

- Potential medical usage

- Harvested human Lung Preservation With the Use of Hibernation Trigger Factors

- First Application In Hibernate Creation a web application using servlet in hibernate

4.5 Indefinite lifespan

Indefinite lifespan (also known as indefinite life extension) is a term used in the life extension movement and transhumanism to refer to the hypothetical longevity of humans (and other life-forms) under conditions in which ageing is effectively and completely prevented and treated. Their lifespans would be "indefinite" (that is, they would not be "immortal"), because protection from the effects of aging on health does not guarantee survival. Such individuals would still be susceptible to accidental or intentional death by disease, starvation, physical trauma, and so on, but not death from aging. Semantically, "indefinite lifespan" is more accurate than "immortality" which, especially in religious contexts, implies an inability to die.

4.5.1 Longevity escape velocity

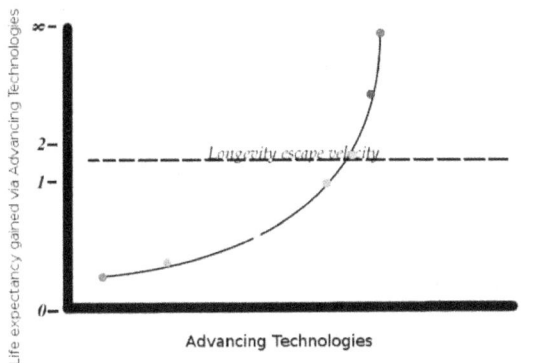

"*The first 1000-year-old is probably only ~10 years younger than the first 150-year-old*"

Main article: Longevity escape velocity

Longevity escape velocity is a term used in the life extension movement. It is a hypothetical situation in which life expectancy is being increased faster than time is being expended. For example, in a given year in which longevity escape velocity would be maintained, technological advances would increase life expectancy more than that year took away by passing by.

4.5.2 Not immortality

The terms "immortality" and "eternal youth" are often used as synonyms for "indefinite lifespan", but they carry connotations from their other contexts which science has deemed to be impossible. That is, *immortal* means "incapable of dying". *Eternal* implies guaranteed existence for eternity, and in this context is also implausible because of entropy. Even if cures were found for all the degenerative diseases, and effective treatments were developed for all the processes of aging, so that bodies could be maintained as easily as cars can be repaired, people would still be killed in accidents, slain in wars, choose to die, etc.

The term indefinite lifespan represents a more achievable state of affairs, because it merely implies freedom from death by aging or infirmity.

The use of the term is also sometimes favored for reasons of linguistic aesthetics, in the same way that the term birth control is preferred to "*birth prevention*" or "*birth elimination*" which both imply, as does 'immortality', that the choice is one-time only and has permanent consequences. Whereas the point of 'indefinite lifespan', like the point of 'birth control', is to gain the opportunity to lead one's life in a more conscious and deliberate manner.

4.5.3 Probability

This question is twofold. On the one hand it can be interpreted to mean, "Will a cure (or program of effective treatments) for aging ever be developed?" while on the other hand it could mean "Will the effective treatment of aging become available soon enough for those alive today to take advantage of it?" The answer to the first question is conditional on medical advancement: if medical science continues to advance in the fields of biogerontology and bioengineering, then some people hope the answer is "yes, that it will happen eventually, excepting if some event or series of events were to prevent the further advance of biological science" (*see* Risks to civilization, humans and planet Earth *and the* Doomsday Clock). Many scientists researching this area at the moment do not agree. They see a problem in not just individual diseases but in failure of repair mechanisms alluded to above in the discussion of thermodynamic considerations.

While science is constantly advancing and technology is becoming ever more sophisticated, the human body and mind are finitely complex and have not changed significantly in one hundred thousand years, and the aging process has not, in that time, become any more damaging (*which, in short, is why we live three times as long on average in the twenty-first century as we did ten thousand years before*).[1]

The answer to the second question depends on two factors: the first being *how fast* medical science advances, and the second being *how well* each person takes care of himself (such as utilizing the best available life extension technology or not, and generally eating and behaving in a healthful and non-degrading way), both of which may affect whether or not a given person is still alive when the cure (or set of treatments) becomes available. This strategy is captured in the subtitle "Live Long Enough to Live Forever" of the life extension book *Fantastic Voyage*, by Ray Kurzweil and Terry Grossman.

The second factor to the second question hinges on the first factor - no amount of healthy living will enable somebody alive today to reach the point of indefinite lifespan if medical science is curtailed significantly, or if aging turns out to be massively more complex than currently believed. However, if biomedical gerontology continues to improve, if somatic genetic engineering becomes safe and effective (and is not banned by opponents) within the relatively near future, it may be conceivable for some of those now alive to attain indefinite lifespans.

According to biogerontologist Marios Kyriazis, indefinite lifespans will become possible (even inevitable) because of inherent properties of natural laws governing human evolution.[2] Kyriazis believes that[3] as humanity is enhanced by technology, human evolution by natural selection will become redundant, and humans will continue to evolve through an indefinitely-long process of self-development. This process will necessitate the elimination of death due to aging.[4]

4.5.4 Proposed techniques

Main article: Life extension

Strategies for Engineered Negligible Senescence is a proposed research program for repairing all types of age-related damage.

Calorie restriction has been presented as a piece of the puzzle of reaching actuarial escape velocity.[5][6][7] Other proposed techniques include genetic engineering, telomere extension, organ regeneration, nanotechnology, and even mind uploading.[8]

On the theory that the primary cause of aging is DNA damage [see DNA damage theory of aging and DNA damage (naturally occurring)], there are, in principle, two ways of reducing DNA damage in cells, and thus allowing indefinite extension of lifespan. These are:

1. Preventing the occurrence of DNA damage, and

2. Repairing the DNA damage after it has occurred.

There is a large body of literature on antioxidant phytochemicals that reduce the occurrence of oxidative DNA damage. However, when intervention trials were carried out using these antioxidants as dietary supplements and cancer as the endpoint, the results generally proved disappointing.[9][10]

On the other hand, there seems to be evidence that certain dietary components stimulate repair of DNA damage, and protect against cancer as an endpoint. One of these is chlorogenic acid, a major component of, and absorbable from, coffee.[11] Coffee is protective against colorectal cancer,[12] and chlorogenic acid and its metabolites increase the protein expression levels of two DNA repair enzymes: Pms2 and PARP.[13] Another compound that protects against the early stages of cancer is naringenin, a citrus flavonoid.[14] Naringenin was shown to increase the mRNA expression levels of two DNA repair enzymes, DNA pol beta and OGG1.[15]

4.5.5 See also

- Aging

- Anti-aging movement

- Biological immortality

- DNA damage theory of aging

- Dyson's eternal intelligence

- Eternal youth

- HeLa

- Howard Families

- Immortality

- Life extension

- List of life extension related topics

- Longevity

- Maximum life span

- Nootropics

- Senescence

- Technological singularity

4.5.6 References

[1] Cairns J (1997). "Matters of Life and Death" Princeton University Press, Princeton, N.J. (see pages 8-13) ISBN 9780691002507

[2] "The ELPIs Theory - The ELPIs Theory". Elpistheory.info. Retrieved 2011-10-04.

[3] "Immortality". Immortalhumans.com. 2011-06-24. Retrieved 2011-10-04.

[4] https://acrobat.com/#d=MAgyT1rkdwono-lQL6thBQ

[5] Traister, Rebecca (November 22, 2006), "Diet your way to a long, miserable life!", *Salon.com*, retrieved 2008-10-31

[6] Dibbell, Julian (October 23, 2006), "The Fast Supper", *New York Magazine*

[7] Birnbaum, Ben (2006), "Extension program", *Boston College Magazine*

[8] "TIME". TIME. Retrieved 2011-10-04.

[9] Collins AR (2005). Antioxidant intervention as a route to cancer prevention. European Journal of Cancer 41:1923-1930. PMID 16111883

[10] Williams CD (2013). Antioxidants and prevention of gastrointestinal cancers. Curr Opin Gastroenterol 29(2):195-200. doi: 10.1097/MOG.0b013e32835c9d1b. PMID 23274317

[11] Del Rio D, Stalmach A, Calani L, Crozier A (2010). Bioavailability of coffee chlorogenic acids and green tea flavan-3-ols. Nutrients 2(8):820-33. doi: 10.3390/nu2080820. PMID 22254058

[12] Li G, Ma D, Zhang Y, Zheng W, Wang P (2012). Coffee consumption and risk of colorectal cancer: a meta-analysis of observational studies. Public Health Nutr 16(2):346-357. doi: 10.1017/S1368980012002601. PMID 22694939

[13] Bernstein H, Crowley-Skillicorn C, Bernstein C, Payne CM, Dvorak K, Garewal H (2007). Dietary compounds that enhance DNA repair and their relevance to cancer and aging. In "New Research on DNA Repair" (Breehn R. Landseer, editor), Chapter IV, 99-113. ISBN 978-1-60021-385-4 https://www.novapublishers.com/catalog/product_info.php?products_id=43814

[14] Leonardi T, Vanamala J, Taddeo SS, Davidson LA, Murphy ME, Patil BS, Wang N, Carroll RJ, Chapkin RS, Lupton JR, Turner ND (2010). Apigenin and naringenin suppress colon carcinogenesis through the aberrant crypt stage in azoxymethane-treated rats. Exp Biol Med (Maywood) 235(6):710-7. doi: 10.1258/ebm.2010.009359 PMID 20511675

[15] Gao K, Henning SM, Niu Y, Youssefian AA, Seeram NP, Xu A, Heber D (2006). The citrus flavonoid naringenin stimulates DNA repair in prostate cancer cells. J Nutr Biochem 17(2):89-95. PMID 16111881

4.5.7 Further reading

1. *Fantastic Voyage: The Science Behind Radical Life Extension* Raymond Kurzweil and Terry Grossman M.D., Rodale. 2004. 452pp. ISBN 1-57954-954-3

2. Fahy GM; Wowk B; Wu Jun, P; Rasch C; et al. (2004), Cryopreservation of organs by vitrification: perspectives and recent advances, Cryobiology, Vol.48,Iss.2;p. 157

3. Blackstone, E. Morrison, M. and Roth, M. (2005) Hydrogen Sulfide Induces a Suspended Animation-like State in Mice, Science, Vol. 308, page 518.

4.6 Supercooling

Not to be confused with Superfluidity or Subcooling.

Supercooling, also known as **undercooling**,[1] is the process of lowering the temperature of a liquid or a gas below its freezing point without it becoming a solid.

4.6.1 Explanation

A liquid crossing its standard freezing point will crystalize in the presence of a seed crystal or nucleus around which a crystal structure can form creating a solid. Lacking any such nuclei, the liquid phase can be maintained all the way down to the temperature at which crystal homogeneous nucleation occurs. Homogeneous nucleation can occur above the glass transition temperature, but if homogeneous nucleation has not occurred above that temperature an amorphous (noncrystalline) solid will form.

Water normally freezes at 273.15 K (0 °C or 32 °F) but it can be "supercooled" at standard pressure down to its crystal homogeneous nucleation at almost 224.8 K (−48.3 °C/−55 °F).[2][3] The process of supercooling requires that water be pure and free of nucleation sites, which can be achieved by processes like reverse osmosis, but the cooling itself does not require any specialised technique. If water is cooled at a rate on the order of 10^6 K/s, the crystal nucleation can be avoided and water becomes a glass. Its glass transition temperature is much colder and harder to determine, but studies estimate it at about 136 K (−137 °C/−215 °F).[4] Glassy water can be heated up to approximately 150 K (−123 °C/−189.4 °F) without nucleation occurring.[3] In the range of temperatures between 231 K (−42 °C/−43.6 °F) and 150 K (−123 °C/−189.4 °F) experiments find only crystal ice.

Droplets of supercooled water often exist in stratiform and cumulus clouds. Aircraft flying through these clouds see an abrupt crystallization of these droplets, which can result in the formation of ice on the aircraft's wings or blockage of its instruments and probes, unless the aircraft are equipped with an appropriate de-icing system. Freezing rain is also caused by supercooled droplets.

The process opposite to supercooling, the melting of a solid above the freezing point, is much more difficult, and a solid will almost always melt at the same temperature for a given pressure. For this reason, it is the melting point which is usually identified, using melting point apparatus; even when the subject of a paper is "freezing-point determination", the actual methodology is "the principle of observing the disappearance rather than the formation of ice".[5] It is possible, at a given pressure, to superheat a liquid above its boiling point without it becoming gaseous.

Supercooling is often confused with freezing-point depression. Supercooling is the cooling of a liquid below its freezing point without it becoming solid. Freezing point depression is when a solution can be cooled below the freezing point of the corresponding pure liquid due to the presence of the solute; an example of this is the freezing point depression that occurs when sodium chloride is added to pure water.

4.6.2 Constitutional supercooling

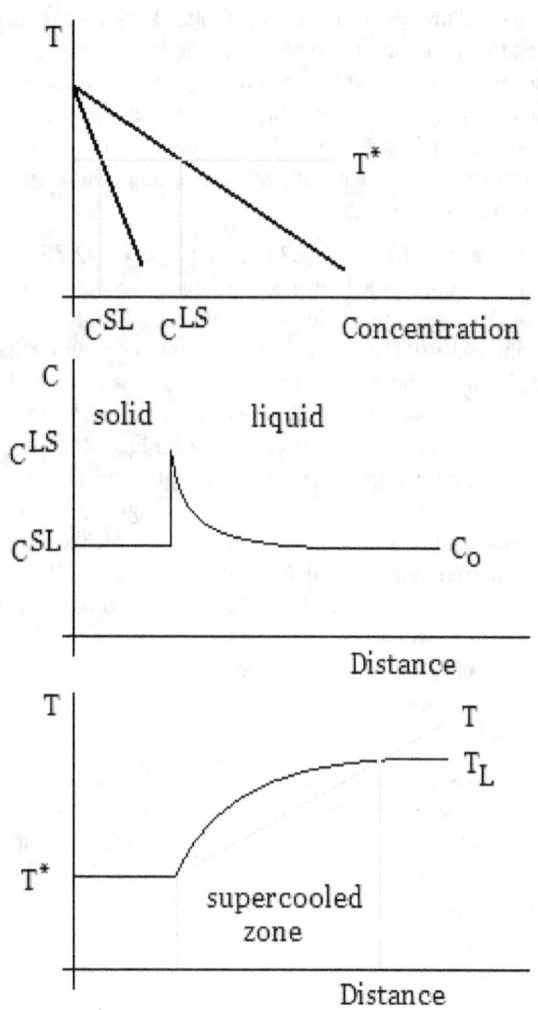

Constitutional supercooling – phase diagram, concentration, and temperature

Constitutional supercooling, which occurs during solidification, is due to compositional changes, and results in cooling a liquid below the freezing point ahead of the solid–liquid interface. When solidifying a liquid, the interface is often unstable, and the velocity of the solid–liquid interface must be small in order to avoid constitutional supercooling.

Supercooled zones are observed when the liquidus temperature gradient at the interface is larger than the temperature

gradient.

$$\left.\frac{\partial T_L}{\partial x}\right|_{x=0} > \frac{\partial T}{\partial x}$$

or

$$m \left.\frac{\partial C_L}{\partial x}\right|_{x=0} > \frac{\partial T}{\partial x}$$

The slope of the liquidus phase boundary on the phase diagram is $m = \partial T_L / \partial C_L$

The concentration gradient is related to points, C^{LS} and C^{SL}, on the phase diagram:

$$\left.\frac{\partial C_L}{\partial x}\right|_{x=0} = -\frac{C^{LS} - C^{SL}}{D/v}$$

For steady-state growth $C^{SL} = C_0$ and the partition function $k = \frac{C^{SL}}{C^{LS}}$ can be assumed to be constant. Therefore the minimum thermal gradient necessary to create a stable solid front is as expressed below.

$$\frac{\partial T}{\partial x} < \frac{m C_0 (1 - k) v}{k D}$$

For more information, see the equation (3) of [6]

4.6.3 In animals

In order to survive extreme low temperatures in certain environments, some animals undergo forms of supercooling that allow them to remain unfrozen and avoid cell damage and death. There are many techniques that aid in supercooling, such as the production of antifreeze proteins, or AFPs, which bind to ice crystals to prevent water molecules from binding and spreading the growth of ice.[7] The winter flounder is one such fish that utilizes these proteins to survive in its frigid environment. Noncolligative proteins are secreted by the liver into the bloodstream.[8] Other animals use colligative antifreezes, which increases the concentration of solutes in their bodily fluids, thus lowering their freezing point. Fish that rely on supercooling for survival must also live well below the water surface, because if they came into contact with ice nuclei they would freeze immediately. Animals that undergo supercooling to survive must also remove ice-nucleating agents from their bodies because they act as a starting point for freezing. Supercooling is also common in insects, reptiles, and other ectotherms, with insects being able to survive in the coldest environments out of any supercooling animals.

It should be noted that supercooling is a last resort for animals. The best option is to move to a warmer environment if possible. As an animal gets farther and farther below its original freezing point the chance of spontaneous freezing increases dramatically for its internal fluids, as this is a thermodynamically unstable state. The fluids eventually reach the supercooling point, which is the temperature at which the supercooled solution freezes spontaneously due to being so far below its normal freezing point.[9] Animals unintentionally undergo supercooling and are only able to decrease the odds of freezing once supercooled. Even though supercooling is essential for survival, there are many risks associated with it.

4.6.4 Applications

One commercial application of supercooling is in refrigeration. Freezers can cool drinks to a supercooled level[10] so that when they are opened, they form a slush. Another example is a product that can supercool the beverage in a conventional freezer.[11] The Coca-Cola Company briefly marketed special vending machines containing Sprite in the UK, and Coke in Singapore, which stored the bottles in a supercooled state so that their content would turn to slush upon opening.[12]

Supercooling was successfully applied to organ preservation by the group of Martin Yarmush and Korkut Uygun in the Center for Engineering in Medicine at Massachusetts General Hospital/Harvard Medical School. Livers that were later transplanted into recipient animals were preserved by supercooling for up to 96 hours, quadrupling the limits of what could be achieved by conventional liver preservation methods. The livers were supercooled to a temperature of −6°C in a specialized solution that protected against freezing and injury from the cold temperature.[13]

Another potential application is drug delivery. In 2015 researchers demonstrated the ability to crystallize membranes at a specific time. Liquid-encapsulated drugs can potentially be delivered to the site and with a slight environmental change, the liquid rapidly changes into a crystalline form that releases the drug.[14]

4.6.5 See also

- Amorphous solid

- Pumpable ice technology

- Subcooling

- Ultracold atom

- Viscous liquid

- Freezing rain

4.6.6 References

[1] Rathz, Tom. "Undercooling". NASA. Archived from the original on 2010-01-12. Retrieved 2010-01-12.

[2] Moore, Emily; Valeria Molinero (24 November 2011). "structural transformation in supercooled water controls the crystallization rate of ice". *Nature* **479**: 506–508. arXiv:1107.1622. Bibcode:2011Natur.479..506M. doi:10.1038/nature10586. Retrieved 24 November 2011.

[3] Debenedetti & Stanley 2003, p. 42

[4] Insights into Phases of Liquid Water from Study of Its Unusual Glass-Forming Properties, C. Austen Angell, Science 319, 582 (2008); .

[5] "A new method of freezing-point determination for small quantities", J. A. Ramsay, *J. Exp. Biol.*.1949; 26

[6] page from 99~100

[7] J.G. Duman (2001). "Antifreeze and ice nucleator proteins in terrestrial arthropods". *Annual Review of Physiology* **63**: 327–357. doi:10.1146/annurev.physiol.63.1.327. PMID 11181959.

[8] Garth L Fletcher, Choy L Hew, and Peter L Davies (2001). "Antifreeze Proteins of Teleost Fishes". *Annual Review of Physiology* **63**: 359–390. doi:10.1146/annurev.physiol.63.1.359. PMID 11181960.

[9] C.H. Lowe, P.J. Lardner, and E.A. Halpern (1971). "Supercooling in reptiles and other vertebrates". *Comparative Biochemistry and Physiology* **39A** (1): 125–135. doi:10.1016/0300-9629(71)90352-5. PMID 4399229.

[10] Archived March 1, 2009 at the Wayback Machine

[11] Archived January 23, 2010 at the Wayback Machine

[12] Charlie Sorrel (2007-09-21). "Coca Cola Plans High Tech, Super Cool Sprite". *Wired*. Condé Nast. Retrieved 2013-12-05.

[13] Berendsen, TA; Bruinsma, BG; Puts, CF; Saeidi, N; Usta, OB; Uygun, BE; Izamis, Maria-Louisa; Toner, Mehmet; Yarmush, Martin L; Uygun, Korkut (2014). "Supercooling enables long-term transplantation survival following 4 days of liver preservation". *Nature Medecine* **20** (7): 790–3. doi:10.1038/nm.3588.

[14] Hunka, George (2015-05-06). "A "super cool" way to deliver drugs". R&D. Retrieved May 2015.

4.6.7 Further reading

- Debenedetti, P. G.; Stanley, H. E. (2003). "Supercooled and Glassy Water" (PDF). *Physics Today* **56** (6): 40–46. Bibcode:2003PhT....56f..40D. doi:10.1063/1.1595053.

- Giovambattista, N.; Angell, C. A.; Sciortino, F.; Stanley, H. E. (July 2004). "Glass-Transition Temperature of Water: A Simulation Study" (PDF). *Physical Review Letters* **93** (4): 047801. arXiv:cond-mat/0403133. Bibcode:2004PhRvL..93d7801G. doi:10.1103/PhysRevLett.93.047801. PMID 15323794.

- Rogerson, M. A.; Cardoso, S. S. S. (April 2004). "Solidification in heat packs: III. Metallic trigger". *AIChE Journal* **49** (2): 522–529. doi:10.1002/aic.690490222.

4.6.8 External links

- Supercooled water and coke on YouTube

- Supercooled water on YouTube

- Super Cooled Water #2 on YouTube

- Supercooled Water Nucleation Experiments on YouTube

- Supercooled liquids on arxiv.org

- Radiolab podcast on supercooling

Chapter 5

Text and image sources, contributors, and licenses

5.1 Text

- **Cryonics** *Source:* https://en.wikipedia.org/wiki/Cryonics?oldid=689748336 *Contributors:* Bryan Derksen, The Anome, Gianfranco, SimonP, Maury Markowitz, Chuq, Edward, Ubiquity, Patrick, JohnOwens, Gabbe, Ixfd64, Paul A, Ams80, AllanR~enwiki, Ciphergoth, Cryoboy, Timwi, Dysprosia, Tempshill, Omegatron, PeterMerel, Mignon~enwiki, Scott Sanchez, Pakaran, Jerzy, Finlay McWalter, Huangdi, Drxeno-cide, Paranoid, Munin, Naddy, TimR, Jondel, Pablo-flores, Juhaz, David Gerard, Buster2058, DocWatson42, Mporter, Bkonrad, Dratman, Fjarlq, Dmmaus, Bobblewik, Christopherlin, Sonjaaa, Marcus Beyer, Loremaster, Rdsmith4, Mike Rosoft, Discospinster, Brianhe, Rich Farm-brough, Plumpy, Tsujigiri~enwiki, Smalljim, Bollar, Drw25, Pharos, Jason One, Guy Harris, Philip Cross, NamfFohyr, Lightdarkness, Sligocki, Stack, Hu, Pauli133, Kitch, Btornado, Simetrical, Woohookitty, FeanorStar7, Benbest, JeremyA, M412k, Male1979, Spyros Pantenas, Pic-tureuploader, Bjtitus, EggplantWizard, Ashmoo, BD2412, Bikeable, Rjwilmsi, Hiberniantears, JoshuacUK, Stardust8212, X1011, NeonMerlin, The wub, Ttwaring, Tbone, FlaBot, SchuminWeb, Windchaser, Q11, Nivix, AI, RexNL, Mattman00000, Chobot, Karch, Bgwhite, Check two you, Roboto de Ajvol, YurikBot, Wavelength, Mikalra, Snappy, MattWright, RussBot, Pigman, Chris Capoccia, Rodasmith, Gaius Cor-nelius, Draeco, Thechosenone021, Dialectric, Muppetmaster, ONEder Boy, Howcheng, Chrisbrl88, Rmky87, Tony1, Occono, Cardsplayer4life, Haemo, BlueZenith, Mike Dillon, Dspradau, BorgQueen, GraemeL, SigmaEpsilon, 4shizzal, Mynabull, KnightRider~enwiki, SmackBot, Ob-have, Reedy, Postbagboy, InverseHypercube, JohnPomeranz, McGeddon, Ankoch, Niayre, Cryobiologist, Arny, Kintetsubuffalo, Septegram, Commander Keane bot, Gilliam, Ohnoitsjamie, Manuelomar2001, Bluebot, Kurykh, Audacity, Shaggorama, Darnoconrad, RDBrown, Crash-matrix, Ikiroid, Sbharris, Scwlong, Valdezlopez, SpinyNorman, Facto, APRCooper, Mr President, WhtRbt, Bejnar, Thor Dockweiler, Pink-tulip, Dane Sorensen, Gloriamarie, Reaganamerican, Kingfish, Aussie Alchemist, Meteshjj, Shadowcaster187, Twile, JoshuaZ, RomanSpa, JHunterJ, MarkSutton, Smith609, Beetstra, Raj89er, Grandpafootsoldier, Xiaphias, Mauro Bieg, Waggers, Kyle C Haight, Nicolasdz, Xionbox, Gavrilov, Al59, Danlev, Tawkerbot2, Cryptic C62, Rabidchipmunk666, Joostvandeputte~enwiki, RSido, CmdrObot, Coolcamxl, Kushal one, Nadyes, TVC 15, Neelix, Gregbard, .Hustler, Kribbeh, Yaris678, Cydebot, JBarletta, Gogo Dodo, Beefnut, Teratornis, GangstaEB, JodyB, Col. Hauler, Thijs!bot, Epbr123, BladeStopper, Fetternity, Headbomb, Peter Deer, Ufwuct, Ettinger, JustAGal, EdgarSwank, AntiVandalBot, MoogleDan, Dr. Blofeld, King ofall1, LibLord, Blacksun1942, Wayiran, JAnDbot, Freepsbane, Albany NY, Ishikawa Minoru, GearedBull, Cryopunk, Noip~enwiki, Cecelia Hensley, Olikea, GirlForLife, DropZone, Bobby H. Heffley, Benjamintchip, Rajpaj, Edward321, Dlempa, Shorelander, Stephenchou0722, MartinBot, Theo75~enwiki, Xsmasher, R'n'B, Nono64, Lilac Soul, Pomte, Aznoble, Hans Dunkelberg, Mikael Häggström, Belovedfreak, Cobi, Darkshadowjedi, Inter16, SoCalSuperEagle, Aagtbdfoua, Arundel77, Deor, VolkovBot, BoogaLouie, Poser-boarder, Despres, Someguy1221, PrinceGloria, Corvus cornix, Teeg82, PDFbot, Enviroboy, Jordansparks, Spartacus106, Sesshomaru, Laval, Doc James, Legoktm, Rypcord, SieBot, Calliopejen1, Sora-Angelo-Strife, Lucasbfrbot, Araignee, Ouizardus, LeadSongDog, Joe3600, Em-peror001, Yeo123, Rlc02d, Lightmouse, Typeform, Anchor Link Bot, Yair rand, Lord Opeth, Chem-awb, Martarius, FlamingSilmaril, Clue-Bot, Samba pa ti, Bellmessenger1, Unbuttered Parsnip, Gaia Octavia Agrippa, Niceguyedc, Phenylalanine, Mandalorian NerfHerder Maceo, Alexbot, MantisEars, Amozoness6, DumZiBoT, RMerkle, BarretB, AlanM1, MR Bass to you, XLinkBot, Will-B, Gnowor, Dthomsen8, Eleven even, MystBot, SelfQ, Lemmey, Cartermassey, JamesNT, Volucer~enwiki, DOI bot, Prestonwalsh, Soundout, Acs1969, Fieldday-sunday, Mac Dreamstate, Looie496, Robbie0630, Download, LaaknorBot, NailPuppy, Tassedethe, Terwilliger44, Tide rolls, Jarble, Legobot, Zhitelew, Yobot, Fraggle81, Enjoydrawn, Amirobot, EnochBethany, Memgab, AnomieBOT, Daniele Pugliesi, Mugabenator3, Yachtsman1, Citation bot, Que-bec99, LilHelpa, FreeRangeFrog, Mrc1028, Gap9551, Mechfish, Ruy Pugliesi, GrouchoBot, Kalbasa, Shadowjams, Eugene-elgato, Blaueziege, Erik9, FrescoBot, B0gger, Citation bot 1, Redrose64, Milan studio, Double sharp, Sandegud, Phaffo, Lotje, Vrenator, Jtjohnstone, Sgt. R.K. Blue, Kristiani95, Jeffrd10, PleaseStand, RjwilmsiBot, Likemonkeys, Yangosplat222, CalicoCatLover, EmausBot, Acather96, WikitanvirBot, Super48paul, K6ka, XJ90, Dolovis, Wingman4l7, Misterhistory, Ch4OSm4n, Robert Ettinger, Ihardlythinkso, Whoop whoop pull up, Petrb, Will Beback Auto, ClueBot NG, Jack Greenmaven, CocuBot, Jangnathan, SSSheridan, Wikigoodnews, MiguelGMC, H3iu.87xW.k44r.0H3d, Widr, Neoprometheus, Helpful Pixie Bot, BG19bot, Virtualerian, Ahura21, Mark Arsten, Fredirib, Tdwillson, Yelance, MrBill3, Colbert Sesanker, Octopus5650, BattyBot, ChrisGualtieri, Webclient101, Kman216, Mogism, Kbog, Kolinzo, TwoTwoHello, Cupco, DanielTom, Corn cheese, Blacknight87, Randykitty, Harleywijey123, 2010 SO16, Hujraadjohaansen, Fragnetto, Everymorning, Rolf h nelson, CensoredScribe, Saucer-balls, Paupatto, Linuxjava, Monkbot, Lampshade13271, Litlessss, Logibohr, Лайземир, Don Kaustav, BonmotNorth, ComicsAreJustAllRight, LissianaElena89, Gazingpenguin, DiscoTent and Anonymous: 516

- **Cryopreservation** *Source:* https://en.wikipedia.org/wiki/Cryopreservation?oldid=687852500 *Contributors:* Magnus Manske, Bryan Derksen, SimonP, Anthere, Hfastedge, Edward, Gabbe, Rambot, Notheruser, Zoicon5, Samsara, Topbanana, Jerzy, Robbot, David Gerard, Graeme Bartlett, DavidCary, Peruvianllama, Fjarlq, Gadfium, Deglr6328, Discospinster, Guanabot, NeuronExMachina, Vsmith, Mani1, Blade Hirato~enwiki, PrometheusOne, Pt, Art LaPella, Alansohn, Arthena, Hipocrite, Danhash, Dirac1933, Gene Nygaard, Benbest, Marudubshinki, Graham87, Rjwilmsi, Alexjohnc3, Thumbofodin, King of Hearts, YurikBot, Wavelength, Huw Powell, Gaius Cornelius, Shaddack, Rsrikanth05, NawlinWiki, Brian Crawford, Daniel Mietchen, Wknight94, Ninly, Modify, GraemeL, SigmaEpsilon, Arielle Rose, Snalwibma, SmackBot, Cryobiologist, Kurykh, Thumperward, Deli nk, Wikipediatrix, Juandev, Rrburke, VegaDark, Ultraexactzz, ArglebargleIV, Gloriamarie, Arunrajmohan, Mgiganteus1, Hu12, DonaldJB, Switchercat, Wafulz, Dgw, WeggeBot, Almighty Rajah, Uruiamme, AntiVandalBot, Blacksun1942, Txomin, Magioladitis, Andrucius, QuizzicalBee, GirlForLife, XMog, Gwern, Atulsnischal, Rahulpm, R'n'B, J.delanoy, Shawn in Montreal, Mikael Häggström, Sabisteb, Tagus, Jakedimare, Corvus cornix, Mineys, Thisismyrofl, Doc James, Ynagashima, Laoris, Leriredufils, Wizzard2k, Xvani, Chem-awb, ClueBot, Mild Bill Hiccup, Polyamorph, NuclearWarfare, Mlaffs, Wnt, Ost316, M2o6n0k2e7y3, Addbot, DOI bot, Zellfaze, Acs1969, Download, LaaknorBot, Favonian, Rengachen27, HFEA, OlEnglish, SasiSasi, Luckas-bot, Yobot, AnomieBOT, Rubinbot, Bluerasberry, Materialscientist, Spotfixer, Sylwia Ufnalska, FertilityPlan, Qweedsa, ProtectionTaggingBot, Valentino76, FrescoBot, Citation bot 1, I dream of horses, HRoestBot, Haida19, Tom.Reding, Edzmaz, Marie Poise, RjwilmsiBot, TjBot, John of Reading, BLM Platinum, RenamedUser01302013, K6ka, AvicBot, ZéroBot, WaltDisneyFan, Ida Shaw, Liquidmetalrob, H3llBot, Rangoon11, Tallboyt, ClueBot NG, Oddbodz, Bibcode Bot, Virtualerian, Mlthompson76, Altaïr, Thegreatgrabber, BattyBot, Dexbot, Sidelight12, Cupco, Creator123321, VanishedUser 2313214sad1, 2010 SO16, Ijhaniff, DudeWithAFeud, Monkbot, PanosLoizou, RobinNJ37, KasparBot and Anonymous: 145

- **Biostasis** *Source:* https://en.wikipedia.org/wiki/Biostasis?oldid=582336100 *Contributors:* Magnus Manske, Big Bob the Finder, Jni, Bobblewik, Xmnemonic, Klemen Kocjancic, Viriditas, Stemonitis, Benbest, Bluemoose, Wjfox2005, Leptictidium, EncycloPetey, Bluebot, Soap, Alaibot, Wench999, BeefRendang, MiltonT, R'n'B, Colincbn, Unbuttered Parsnip, Legobot, Xqbot, EmausBot, Mohamed-Ahmed-FG, Emil3168 and Anonymous: 9

- **Cryobiology** *Source:* https://en.wikipedia.org/wiki/Cryobiology?oldid=668013080 *Contributors:* JeLuF, SimonP, Lquilter, Stefan-S, Nikai, Robbot, Paranoid, Vespristiano, David Gerard, Falcon Kirtaran, Slowking Man, Marcus Beyer, Kelson, Rich Farmbrough, Vsmith, Mani1, Gilgamesh he, Maurreen, Jumbuck, Wadems, Ceyockey, Nuno Tavares, Woohookitty, Benbest, Bluemoose, Rjwilmsi, Mayumashu, Chobot, Bgwhite, YurikBot, Wavelength, Howcheng, Daniel Mietchen, SmackBot, Cryobiologist, Frap, AP1787, OrphanBot, Drunken Pirate, Gloriamarie, John, IvanLanin, FleetCommand, Cydebot, Skittleys, Headbomb, Drdaveng, AntiVandalBot, Just Chilling, Arch dude, Magioladitis, Ekantik, Cecelia Hensley, GirlForLife, Agricolae, R'n'B, Sputtek, Spellcast, Tameeria, A4bot, Lamro, SieBot, Yerpo, SimonTrew, Riyadi, Chem-awb, Muhends, ClueBot, Prottos007, Ltnemo2000, Sun Creator, Mlaffs, Addbot, DOI bot, Favonian, Luckas-bot, Zhitelew, Yobot, AnakngAraw, JackieBot, Ruimsf, Xqbot, FertilityPlan, Guitpicker07, Dcrjsr, FrescoBot, BenzolBot, Citation bot 1, DrilBot, Fartherred, Tbhotch, EmausBot, Pecanpie123, MithrandirAgain, ClueBot NG, Tideflat, Bart simpson rules, Institut International du Froid, IraChesterfield, Aranea Mortem, MrBill3, ChrisGualtieri, Reatlas, Ashorocetus, Rakeshkacham88, Glaisher, Prokaryotes, Lizia7, Jerodlycett, KasparBot and Anonymous: 53

- **Cryonics Institute** *Source:* https://en.wikipedia.org/wiki/Cryonics_Institute?oldid=686159455 *Contributors:* SimonP, Gabbe, Dcoetzee, Maximus Rex, Mrand, Bearcat, Pretzelpaws, Jfdwolff, Lucky 6.9, Karl Dickman, I9Q79oL78KiL0QTFHgyc, Bsadowski1, Benbest, Rjwilmsi, Imnotminkus, Bgwhite, Malcolma, Deku-shrub, PS2pcGAMER, Sticky Parkin, Slashme, Gloriamarie, CmdrObot, Cydebot, DumbBOT, Richhoncho, RobotG, WinBot, DarthShrine, Blacksun1942, Cecelia Hensley, GirlForLife, R'n'B, Player 03, Antagon, Maproom, Jkeene, Demize, Joe3600, Flyer22 Reborn, Chem-awb, WurmWoode, Alexbot, Adimovk5, SchreiberBike, Edwardmorrill, Dthomsen8, Addbot, DOI bot, Glane23, Lightbot, AnomieBOT, Jim1138, Gap9551, Pmlineditor, Full-date unlinking bot, RjwilmsiBot, GoingBatty, H3llBot, ChuispastonBot, ClueBot NG, H3iu.87xW.k44r.0H3d, Jenvogel, Helpful Pixie Bot, BG19bot, BattyBot, Douglas T Golner, Monkbot, KH-1, Rubbish computer, CIConsultant and Anonymous: 43

- **Cryonics – Freeze Me** *Source:* https://en.wikipedia.org/wiki/Cryonics_%E2%80%93_Freeze_Me?oldid=621875918 *Contributors:* Ciphergoth, Pigman, Ohnoitsjamie, Azumanga1, Willking1979, Yobot, AnomieBOT, TheAMmollusc, Fortdj33, PerlDiver, DASHBot, Rolf h nelson and Anonymous: 1

- **Cryoprotectant** *Source:* https://en.wikipedia.org/wiki/Cryoprotectant?oldid=690199441 *Contributors:* Giraffedata, Benbest, Rjwilmsi, Hulagutten, Nihiltres, Vmenkov, Ashwinr, JHCaufield, Cardsplayer4life, SmackBot, Edgar181, Gilliam, Thumperward, GVnayR, T g7, Kognyto, Beetstra, Nadyes, Cydebot, AntiVandalBot, Magioladitis, GirlForLife, R'n'B, Mikael Häggström, Tameeria, Jvbishop, Lamro, VVVBot, Chem-awb, Mlaffs, DumZiBoT, MystBot, Airplaneman, Addbot, Krasss, Xqbot, FrescoBot, Tom.Reding, RjwilmsiBot, Uploadvirus, ZéroBot, Virtualerian, Thatemooverthere, BattyBot, Comatmebro, Jeremy.winkler, Monkbot and Anonymous: 17

- **Information-theoretic death** *Source:* https://en.wikipedia.org/wiki/Information-theoretic_death?oldid=678219593 *Contributors:* Ciphergoth, Evercat, Charles Matthews, David Gerard, Everyking, Brianhe, Localhost00, Stuartyeates, Simetrical, Mindmatrix, Benbest, Qwertyus, That Guy, From That Show!, SmackBot, Cryobiologist, Bluebot, Sbharris, Bigturtle, Drphilharmonic, Petr Kopač, Rabidchipmunk666, Myncknm, Dkowalski, Tkgd2007, Pwnage8, Flyer22 Reborn, Tesi1700, RMerkle, Addbot, Tornsubject, Bwrs, Jarble, Yobot, AnomieBOT, Gap9551, Crzer07, LucienBOT, Vladimir Shmachkov, ZéroBot, Fredirib, Oracions, Dr Athanasis Boukalis, Lsparrish and Anonymous: 26

- **James Bedford** *Source:* https://en.wikipedia.org/wiki/James_Bedford?oldid=683199295 *Contributors:* Jni, Matt Gies, Mporter, Cam, Roisterer, Ukexpat, Zhmort, Gazpacho, Brianhe, Philip Cross, BDD, Benbest, Rjwilmsi, Srleffler, Conscious, Splash, Kimchi.sg, Emc2, Sparge, Ligulembot, Captainbeefart, Ser Amantio di Nicolao, Gloriamarie, Caejis, Suede~enwiki, Waacstats, JaffaCakeLover, GirlForLife, DGG, R'n'B, Johnpacklambert, Useight, Aymeric78, Ipso2, Guillaume2303, Lightmouse, Chem-awb, ClueBot, WurmWoode, Snocrates, Alexbot, DumZiBoT, Boleyn, RogDel, Anticipation of a New Lover's Arrival, The, Addbot, AVand, Timissimit, Fluffernutter, Dbrandte, OlEnglish, Yobot, Fraggle81, AnomieBOT, Rubinbot, Full-date unlinking bot, RjwilmsiBot, EmausBot, GoingBatty, Philippe277, ZéroBot, Tolly4bolly, ChuispastonBot, ClueBot NG, BattyBot, Kodiologist, VIAFbot, Wikileegh, Babitaarora, TheOtherPhylicia, TaqPol, KasparBot and Anonymous: 44

- **Neuropreservation** *Source:* https://en.wikipedia.org/wiki/Neuropreservation?oldid=677433372 *Contributors:* Frecklefoot, David Gerard, Rich Farmbrough, Benbest, Rjwilmsi, Cardsplayer4life, Cryobiologist, Clicketyclack, Gloriamarie, Jj137, Cecelia Hensley, R'n'B, Player 03, Jordansparks, Niceguyedc, Mlaffs, Htmlcoderexe, RMerkle, Addbot, Luckas-bot, Yobot, AnomieBOT, Xqbot, Trappist the monk, RjwilmsiBot, ZéroBot, Fredirib, Lsparrish and Anonymous: 9

- **Raymond and Monique Martinot** *Source:* https://en.wikipedia.org/wiki/Raymond_and_Monique_Martinot?oldid=685454450 *Contributors:* AdamBradley, Captainmax, Rms125a@hotmail.com, SmackBot, Wizardman, Goodnightmush, Lugnuts, Alaibot, Addere, Mattgirling, Felix

Folio Secundus, Addbot, Yobot, AnomieBOT, Materialscientist, Gap9551, RjwilmsiBot, WikitanvirBot, Frze, Mark Arsten, DoctorKubla, Marcila28, Snehrerc2014 and Anonymous: 1

- **Robert Prehoda** *Source:* https://en.wikipedia.org/wiki/Robert_Prehoda?oldid=621399455 *Contributors:* Auric, Mporter, Koavf, SmackBot, Racklever, Waacstats, Technopat, RjwilmsiBot and Anonymous: 3

- **Shannon Vyff** *Source:* https://en.wikipedia.org/wiki/Shannon_Vyff?oldid=690389064 *Contributors:* David Gerard, Benbest, Mandarax, Rjwilmsi, Nikkimaria, SmackBot, Janthepayne, Edward321, NatGertler, WurmWoode, Addbot, AnomieBOT, Gap9551, RadManCF, FrescoBot, RjwilmsiBot, Benjamin simm, H3llBot, Helpful Pixie Bot, LlamaAl, Decathlete, Monkbot, RobertBDurham, Lsparrish and Anonymous: 2

- **Suspended animation** *Source:* https://en.wikipedia.org/wiki/Suspended_animation?oldid=689458824 *Contributors:* Mav, Michael Hardy, Ixfd64, Ahoerstemeier, Bogdangiusca, Emperorbma, David Latapie, DJ Clayworth, Morwen, Branddobbe, DavidA, Pigsonthewing, Jacek79, Postdlf, Rholton, Litefantastic, Cholling, Meelar, UtherSRG, Mushroom, Cyberia23, David Gerard, Geeoharee, Wolfe, Datepalm17, Mboverload, Kravietz, Bobblewik, LucasVB, Pcarbonn, Kuralyov, Adistav, Gordonjcp, SoM, Bornintheguz, Rich Farmbrough, Mykhal, El C, Viriditas, Andrewpmk, Hu, Ronark, BRW, Pauli133, Kitch, BerserkerBen, Jason Palpatine, Benbest, Jeff3000, Pictureuploader, Obersachse, Rjwilmsi, WCFrancis, Urbane Legend, Trlovejoy, TheDJ, RLent, Wavelength, Quentin X, Kevs, Zafiroblue05, Stephenb, Dumoren, Brasswatchman, Lexicon, Ragesoss, Anetode, Mysid, Gadget850, Pydos, Tequila~enwiki, E Wing, Mr. ATOZ, SmackBot, Radak, Cryobiologist, Gilliam, Tennekis, Chris the speller, Gonzalo84, Thumperward, Deuxhero, Sadads, Oni Ookami Alfador, Esprix, Scwlong, DanMat6288, Pedroshin, Andrea Parton, MeekSaffron, WinstonSmith, Iapetus, Inuyasha20985, VegaDark, Mwtoews, Ohconfucius, Rukario639, Axem Titanium, Shrew, Loodog, Convex hull, Cmh, Dilcoe, Buinne, Sausagerooster, Tjeerdnet, JoeBot, Covenant Elite, CmdrObot, Claytonian, Bradkoch2007, Superoxen, The Photographer, Cydebot, Otto4711, Sonicandtails, Javsav, MaulYoda, Bobblehead, Einmonim, Davidhorman, Viva43, Nick Number, SpongeSebastian, SummerPhD, Bakabaka, Jhsounds, Blacksun1942, Wayiran, DOSGuy, Xhienne, Davewho2, The burnanator, QuantumEngineer, Midnightdreary, TheAllSeeingEye, Khaled Khalil, GirlForLife, DerHexer, Wayne Miller, Infovarius, Ariel., Flo422, Hans Dunkelberg, Ayecee, MrBell, Tdadamemd, ABVS1936, Tobias Willmott, Sangriademuertos, Ricmitch, ABF, James Callahan, TXiKiBoT, Henryodell, Thegoldbar, Jordansparks, Doc James, EdSquareCat, Araignee, Joe3600, Erushford, FearEmbodied, Goustien, Polbot, MixMasterLar, Fratrep, Martarius, ClueBot, Wysprgr2005, Auntof6, Leonard^Bloom, NuclearWarfare, Dana boomer, Classicrockfan42, AncientToaster, Alphador, JohnLM, Chveya, Dowsiewuwu, Addbot, DOI bot, PhearOfTheDark, CanadianLinuxUser, Jorkusmalorkus, Farmercarlos, Lightbot, Phantom in ca, Gambori, Azcolvin429, AnomieBOT, SwiftlyTilt, Lolden, Materialscientist, The High Fin Sperm Whale, Citation bot, Capinedap, Capricorn42, Spartan S58, Samuelsidler, Jayshoot, Tom.Reding, Milan studio, Azmath.rahiman, Erik Evrest, RjwilmsiBot, John of Reading, Haon 2.0, Hll1948, Grondilu, Rangoon11, Terraflorin, Rautamiekka, Wikigoodnews, MerlIwBot, Misc Edit, BG19bot, Virtualerian, Ebmkhfpfclbkjkjpefepoeoiaggoaonh, 42canadian42, MrBill3, BattyBot, Karma61538, Dexbot, Everything Is Numbers, Cupco, Me, Myself, and I are Here, Reatlas, 2010 SO16, YiFeiBot, AssafHod, Monkbot, Lythronaxgestes, Era jade, Lsparrish and Anonymous: 229

- **Alcor Life Extension Foundation** *Source:* https://en.wikipedia.org/wiki/Alcor_Life_Extension_Foundation?oldid=683196999 *Contributors:* William Avery, SimonP, Gabbe, Alfio, BenKovitz, Ehn, Timwi, Omegatron, Bearcat, RedWolf, Hadal, Benc, Michael Snow, Fuelbottle, Jhknight, Jdavidb, Fjarlq, Neilc, Quarl, ScottyBoy900Q, Mindspillage, Brianhe, Rich Farmbrough, Apoc2400, Neilmckillop, Prescott~enwiki, Woohookitty, Deeahbz, Benbest, JFG, GregorB, BD2412, Rjwilmsi, Dimitrii, Nightscream, X1011, Kolbasz, Bi, YurikBot, Malcolma, Dekushrub, Cardsplayer4life, CRKingston, Cryobiologist, Ohnoitsjamie, OAS~enwiki, OOODDD, SpinyNorman, Ccchambers, Gloriamarie, CenozoicEra, Zeppelin462, Thijs!bot, Smjc, Transhumanist, TimVickers, TAnthony, MaxPont, Enoent, Wikidudeman, Cecelia Hensley, GirlForLife, Robwilkes, WarddrBOT, Law Lord, Crazy Canadian, Despres, Pah246, Odysseus1138, Maxim, Jordansparks, Joe3600, Xnatedawgx, KryptoOBERchef, WurmWoode, Niceguyedc, Alexbot, Searcher 1990, Seven twentynine, RMerkle, SelfQ, Addbot, Download, Tassedethe, Lightbot, Luckas-bot, AnomieBOT, Joe2008, Materialscientist, B. Fairbairn, Xqbot, Gap9551, Bookworm2009, Legion23, SUL, Jonesey95, Triplestop, Uberzen1, Wexler's World, Inc., Prozacjack, RjwilmsiBot, DASHBot, GoingBatty, CryoNot, ZéroBot, Yiosie2356, H3llBot, Smith2200, Gone south, H3iu.87xW.k44r.0H3d, Slipstream7, Comfr, BattyBot, Futurist110, Epicgenius, Monkbot, Xoegki, Azealia911 and Anonymous: 95

- **American Cryonics Society** *Source:* https://en.wikipedia.org/wiki/American_Cryonics_Society?oldid=683196044 *Contributors:* Gabbe, Sehrgut, Bearcat, Auric, JesseW, Mboverload, Brianhe, Mailer diablo, Deeahbz, Benbest, Wavelength, RussBot, Malcolma, Deku-shrub, Cardsplayer4life, J-beda, Betacommand, Mairibot, Bluebot, Gloriamarie, CmdrObot, Cydebot, EdgarSwank, Cecelia Hensley, Chem-awb, WurmWoode, Mr.Atoz, Acs1969, Lightbot, Ettrig, Yobot, FrescoBot, Visite fortuitement prolongée, GoingBatty, H3iu.87xW.k44r.0H3d, Neoprometheus, Monkbot and Anonymous: 10

- **Immortalist Society** *Source:* https://en.wikipedia.org/wiki/Immortalist_Society?oldid=671455849 *Contributors:* Gabbe, Geraldshields11, Benbest, Ropcat, Rjwilmsi, RussBot, Deku-shrub, Eagleon, Levineps, Milesgillham, RobotG, TimVickers, Magioladitis, Cecelia Hensley, GirlForLife, Edward321, DGG, R'n'B, Coffee, Dthomsen8, Addbot, Yobot, GrouchoBot, Visite fortuitement prolongée, H3iu.87xW.k44r.0H3d, LlamaAl, Oriole85 and Anonymous: 5

- **KrioRus** *Source:* https://en.wikipedia.org/wiki/KrioRus?oldid=682725040 *Contributors:* Paranoid, Alan Liefting, David Gerard, Rich Farmbrough, Eyu100, Russavia, Wavelength, Deku-shrub, Pegship, Closedmouth, NetRolller 3D, Sbharris, Joostvandeputte~enwiki, Nadyes, Goldenrowley, Edward321, Technopat, Y, Joe3600, Vbond, Snocrates, Addbot, Luckas-bot, Zhitelew, Yobot, Amirobot, AnomieBOT, Gap9551, Legion23, HRoestBot, MastiBot, EmausBot, John of Reading, Antonu, HiW-Bot, ZéroBot, ChuispastonBot, MiguelGMC, Сергей Давыдов, Futurist110, Hmainsbot1, Mathieu.roy.37 and Anonymous: 8

- **Life extension** *Source:* https://en.wikipedia.org/wiki/Life_extension?oldid=690279508 *Contributors:* AxelBoldt, Derek Ross, Taw, Ray Van De Walker, SimonP, Mjb, Michael Hardy, Fred Bauder, Lexor, Gabbe, Paul Benjamin Austin, Gdvorsky, TakuyaMurata, Minesweeper, Ronz, Deisenbe, RickK, David Latapie, Darkhorse, HarryHenryGebel, Ee00224, Altenmann, Stewartadcock, Timrollpickering, Mr-Natural-Health, Fuelbottle, Mshonle~enwiki, DavidCary, ShaneKing, Pretzelpaws, Wolfkeeper, Everyking, Alison, Mark T, Jfdwolff, Fjarlq, ThePhantom, Pgan002, Quarl, PDH, Sam Hocevar, RobertBradbury, Everlong, Xrchz, AAAAA, Discospinster, Rich Farmbrough, ESkog, Filur, Pjf, Lycurgus, Mwanner, Art LaPella, RoyBoy, Neilrieck, Feitclub, Defrosted, Cmdrjameson, Cohesion, Morenus, Scott Ritchie, Audrey, La goutte de pluie, Hooperbloob, Orangemarlin, Alansohn, SlimVirgin, Kurieeto, Hu, Docboat, RJII, TenOfAllTrades, Versageek, Gene Nygaard, Dismas, Woohookitty, Linas, RHaworth, Benbest, Foobie, Dolfrog, SCEhardt, Male1979, Jacj, JohnJohn, EggplantWizard, Ryoung122, Clapaucius, BD2412, Kane5187, Rjwilmsi, Hulagutten, MZMcBride, AI, Gurch, BMF81, Chobot, WillMcC, Bgwhite, Chris Capoccia, SpuriousQ, Lord Jim, Gaius Cornelius, Morphh, Wimt, Pproctor, Nirvana2013, Irishguy, Nick, Kortoso, Cardsplayer4life, Varano, GraemeL,

LeonardoRob0t, Johnpseudo, Kungfuadam, Jonathan.s.kt, Mdwyer, Pankkake, Sarah, SmackBot, Mitteldorf, 1dragon, Verne Equinox, Renesis, Monz, Brossow, Hmusseau, Apers0n, Yamaguchi⯑⯑, Gilliam, Portillo, Marc Kupper, Tyciol, Teemu Ruskeepää, Nativeborncal, Bluebot, Basejumper123~enwiki, Deli nk, Uthbrian, Go for it!, AMK152, Xyzzyplugh, Huon, Nakon, Paul haynes, Pwjb, Richard001, Mistress Selina Kyle, BullRangifer, Metamagician3000, Daniel.Cardenas, Leon..., Tsiehta, DO11.10, Dave Yost, Gobonobo, JoshuaZ, JorisvS, Spintronic, Ekrubntyh, Beetstra, Mr Stephen, Lone wanderer, TastyPoutine, Iridescent, Gavrilov, Joseph Solis in Australia, Aeternus, Exander, Crippled Sloth, Mellindiegirl, Courcelles, Kiwi2795, CmdrObot, Dycedarg, JamesAGreen, Lentower, Penbat, The Enslaver, Lourdess99, Fyrius, Cydebot, Peterdjones, Anthonyhcole, Kernel8008, Trident13, Michael C Price, RXPhd, Richhoncho, Thijs!bot, Simul, Pantothenate, Andyjsmith, Smee, Headbomb, Trevyn, AllenFerguson, MickeyK, Transhumanist, Noclevername, Cultural Freedom, QuiteUnusual, Mack2, Spearman, Pixelface, Wayiran, CNicol, Milonica, The Transhumanist, Borgipedia, Ricks99, VoABot II, Vitaminman, MastCell, Hullaballoo Wolfowitz, Swpb, KCon-Wiki, Dave Muscato, LookingGlass, GirlForLife, Petruspennanen, Jjsimpsn, Wayne Miller, Squidonius, Ekotkie, Saganaki-, DGG, Stenemo, MartinBot, Menscience, Ooga Booga, Hans Dunkelberg, Xris0, Soundfeelings, PeterH2, Rod57, Richcoder, Mikael Häggström, Anonywiki, Tarotcards, Jiu9, Plasticup, Ksy92003, Bonadea, Jkurtzman, 28bytes, TXiKiBoT, Seobeglobal, Mercurywoodrose, Jakedimare, A4bot, Chris hawk, BrianMDelaney, Ask123, Pwestep, Vgranucci, Lfreedom, Adam.J.W.C., Doc James, AlleborgoBot, MunkyJuce69, Stewart222, Spartan, Cyberix, Joe3600, MaynardClark, Donnagates, Nk.sheridan, ThAtSo, AiwaBass, Sunrise, Nancy, Boytaichi, Ward20, Denisarona, Ltnature, ClueBot, GorillaWarfare, Healthwise, The Thing That Should Not Be, Joebloetheschmo, M19ozz5, Uncle Milty, Marainein, Phenylalanine, Muhandes, Sun Creator, Vktri2006, Drdunco, Mlaffs, DumZiBoT, Jytdog, Jovianeye, Zodon, Good Olfactory, Addbot, Jamisonhalliwell, DOI bot, Spfanstiel, Download, Bernstein0275, Keepcalmandcarryon, Thomas Bjørkan, Pinoche, Verbal, Phantom in ca, Karan.102, Yobot, Legobot II, Reenem, AnomieBOT, Stinkypie, Materialscientist, Hartigan08, Citation bot, LilHelpa, Gap9551, Thermoproteus, Crzer07, ⯑⯑⯑, Leylu, Bioethica Americana, Ashershow1, MuffledThud, Aaron Kauppi, Some standardized rigour, Eldudarino, Sko1221, FrescoBot, DoctorDNA, Green06, Whoosit, Abaobab, Henry123ifa, Citation bot 1, Borja-oi1d, Machn, Pinethicket, Triplestop, Mickael Vardo~enwiki, Aboriginal Noise, Wexler's World, Inc., Danaayal, Trappist the monk, Liki von Oppen-Bezalel, Jonkerz, UnderHigh, Reaper Eternal, Diannaa, Minimac, Cmarz, RjwilmsiBot, Fostertom67, Becritical, Joannadt, John of Reading, EME44, Peaceray, ZéroBot, Lyliam, Lovearobot, Ifrank98, Weikit, Wingman417, Aschwole, Drbarrywheeler, Gertie1999, Bstard12, Kleeer, MichaelKovich, Iliastambler, ClueBot NG, Unimags, Wbmag129, Ga2114, Wikigoodnews, Snotbot, Bernie44, Widr, Evenstrepe, Lqsw, Jeffer22, Alice.alison88, Helpful Pixie Bot, Correctingthis, BG19bot, Virtualerian, BruinDukie, Tgdmatters, Moscone, Krinckle, Wiki13, Frze, AMPK, Pramod Vora, JohnWilliams2011, Yelance, MrBill3, Star A Star, BattyBot, Biosthmors, Dknowhow, Nathanielfirst, Dexbot, Decadeologee, A53collins, Jago45, Everything Is Numbers, Dmitry Dzhagarov, Joolzzt, Nils119, Faizan, Randykitty, Mostly Translucent, Namenerd, Carrot Lord, PraetorianFury, Kendram, DanielMarton, BradYard, New worl, Slugrex, Jerry5208, Thevideodrome, Seppi333, Yosarian2, Angusoftheair, Slgonzalez, Kohlins, Rbaughman7, 22merlin, Idanoyes, Monkbot, Bwasserott, AviaEfrat, Oiyarbepsy, Waters.Justin, Kelly120913, Haptic-feedback and Anonymous: 343

- **Life Extension Society** *Source:* https://en.wikipedia.org/wiki/Life_Extension_Society?oldid=665848512 *Contributors:* Bearcat, Benbest, Trlovejoy, SmackBot, Whpq, Gloriamarie, JamesBWatson, Edward321, DGG, Emeraude, Yobot, Milan studio, H3iu.87xW.k44r.0H3d, Filizeki and Anonymous: 4

- **Suspended Animation, Inc** *Source:* https://en.wikipedia.org/wiki/Suspended_Animation%2C_Inc?oldid=675388227 *Contributors:* Bearcat, David Gerard, Graeme Bartlett, Benbest, Bgwhite, Dream out loud, Edward321, Aboutmovies, JL-Bot, Mhockey, Addbot, Hairhorn, ChuispastonBot, ClueBot NG, Jenvogel, Lsparrish and Anonymous: 1

- **21st Century Medicine** *Source:* https://en.wikipedia.org/wiki/21st_Century_Medicine?oldid=682004105 *Contributors:* Christopher Mahan, David Gerard, Bobblewik, Marcus Beyer, Rich Farmbrough, Alistair1978, Aecis, AllyUnion, Froggy~enwiki, Ceyockey, Woohookitty, Benbest, Graham87, Rjwilmsi, Efficacy, Huw Powell, Banana04131, Hmains, Bluebot, BullRangifer, TPO-bot, Blacksun1942, Beaumont, R'n'B, Fences and windows, Addbot, Luckas-bot, RjwilmsiBot, DASHBot, H3llBot, Lsparrish and Anonymous: 5

- **Chemical brain preservation** *Source:* https://en.wikipedia.org/wiki/Chemical_brain_preservation?oldid=671098564 *Contributors:* Stone, David Gerard, Mkosmul, Benbest, SmackBot, Jakedimare, Jordansparks, Malcolmxl5, Flyer22 Reborn, Mild Bill Hiccup, Boing! said Zebedee, CWatchman, Vejvančický, Yobot, PelloPelette, Yiosie2356, Flax5, Conifer and Anonymous: 9

- **Extropianism** *Source:* https://en.wikipedia.org/wiki/Extropianism?oldid=684778088 *Contributors:* Eloquence, Bryan Derksen, Arvindn, Michael Hardy, Alfio, Sir Paul, Vroman, Pedant17, VeryVerily, Intangir, Perl, Enkrates, Eequor, Bobblewik, Antandrus, Loremaster, Tsemii, Kate, Mindspillage, Brianhe, Matthewfallshaw, Kaszeta, Mjk2357, Schaefer, Geraldshields11, Dduane, Benbest, JFG, Toussaint, Rjwilmsi, FlaBot, GangofOne, JoeMystical, KSchutte, StephenWeber, Deodar~enwiki, SmackBot, Jim62sch, Declare, Cryobiologist, Hmains, Deli nk, Tamfang, MHoerich, Dacoutts, Metamagician3000, Robofish, Mattpersons, Ckatz, Metavalent, Zaz-en, Natasha Vita-More, Ken Gallager, Gregbard, MikeWren, Cydebot, Michael C Price, Noclevername, AntiVandalBot, The Transhumanist, Seckelberry1, Edward321, R'n'B, Aswarp, Jeepday, Tarotcards, DadaNeem, AlnoktaBOT, Dinobobicus, Cnilep, EverGreg, Nstanosheck, Jwclement, JJRhetorical, Tnxman307, Xme, Addbot, Lightbot, Luckas-bot, Yobot, Cosmicv, OpenFuture, Citation bot, YoungTesla, Guido.willemsen, FrescoBot, LucienBOT, Mindosis, Machine Elf 1735, Qliphoth, Lotje, Brichard37, EmausBot, ZéroBot, Wingman417, David van bruwaene, PatrickJmahoney, Buddha Manchester, Colbert Sesanker, ChrisGualtieri, Khazar2, IjonTichyIjonTichy, Jrupe.gmu, Monkbot, Extropianu, Aliensyntax and Anonymous: 56

- **Hibernation** *Source:* https://en.wikipedia.org/wiki/Hibernation?oldid=690323053 *Contributors:* Vicki Rosenzweig, Andre Engels, SimonP, TerrapinDundee, Hephaestos, Cprompt, Lexor, DavidWBrooks, Jimfbleak, Mxn, Tempshill, Mperkins, Wetman, Shantavira, Bearcat, Robert-Michel~enwiki, Zandperl, Postdlf, Cevally, ShArky, UtherSRG, Robinh, Xanzzibar, Juhaz, MPF, Everyking, Fleminra, Niteowlneils, Revth, Jackol, OldakQuill, Zeimusu, Robert Brockway, DragonflySixtyseven, GeoGreg, WpZurp, Peter bertok, Grm wnr, Cacycle, Alistair1978, Andrejj, Swid, Beska, Plugwash, Ylee, CanisRufus, Shanes, Bobo192, Viriditas, Scheming Eyebrows, Mithent, Halsteadk, RobertStar20, Fritzpoll, Velella, Vuo, Tsuba~enwiki, Blaxthos, Gmaxwell, Angr, Evolve75, ChrisNoe, Halfacanuck, Xaliqen, Pdn~enwiki, MiG, Isnow, Mandarax, Graham87, A Train, Rjwilmsi, Wikifier, FlaBot, VKokielov, Nihiltres, Vsion, RexNL, Shao, Chobot, Korg, Dj Capricorn, YurikBot, Whoisjohngalt, Mikalra, Hairy Dude, Pacaro, Hede2000, DE, Stephenb, Gaius Cornelius, Nicke L, AJHalliwell, Voyevoda, Korny O'Near, The Obfuscator, Decapod73, Scs, Elizabeyth, Ms2ger, Deeday-UK, LeonardoRob0t, Jonathan.s.kt, Some guy, Steve G~enwiki, Yakudza, SmackBot, Pwt898, Snowmobile, ScaldingHotSoup, Eskimbot, Gilliam, Andy M. Wang, Kurykh, Thumperward, Salvor, Wine Guy, Answerthis, Smooth O, Bowlhover, Astroview120mm, Vina-iwbot~enwiki, Kukini, TenPoundHammer, SS2005, Mjpage, J 1982, Hqduong, Mgiganteus1, HisSpaceResearch, Iridescent, IvanLanin, Sue in az, Tawkerbot2, Ebonyraven, Ale jrb, Dynzmoar, Jsmaye, Hi There, Erasmussen, Herd of Swine, Dominicanpapi82, Gogo Dodo, Rmcgiv, Javsav, Dyanega, הרפסב, SuperJ587, Mojo Hand, Headbomb, Sobreira, Marek69, JSmith60, Ufwuct, Davidhorman, Escarbot, AntiVandalBot, Luna Santin, DarkAudit, Zebedeezbd, Chill doubt, JAnDbot, NapoliRoma, MER-C, Quentar~enwiki, Hedjohnston,

5.2 Images

5.3   Content license

www.ingramcontent.com/pod-product-compliance
Lightning Source LLC
Chambersburg PA
CBHW081559170526
45166CB00009B/2750